Sustainable Values, Sustainable Change

Sustainable Values, Sustainable Change

A GUIDE TO ENVIRONMENTAL DECISION MAKING

Bryan G. Norton

THE UNIVERSITY OF CHICAGO PRESS | CHICAGO AND LONDON

Bryan G. Norton is Distinguished Professor Emeritus of philosophy and environmental policy in the School of Public Policy at the Georgia Institute of Technology. He is the author or editor of several books, including, most recently, *Sustainability: A Philosophy of Adaptive Ecosystem Management*, also published by the University of Chicago Press.

The University of Chicago Press, Chicago 60637
The University of Chicago Press, Ltd., London
© 2015 by The University of Chicago
All rights reserved. Published 2015.
Printed in the United States of America

24 23 22 21 20 19 18 17 16 15   1 2 3 4 5

ISBN-13: 978-0-226-19731-9 (cloth)
ISBN-13: 978-0-226-19745-6 (paper)
ISBN-13: 978-0-226-19759-3 (e-book)
DOI: 10.7208/chicago/ 9780226197593.001.0001

Library of Congress Cataloging-in-Publication Data

Norton, Bryan G., author.
    Sustainable values, sustainable change : a guide to environmental decision making / Bryan G. Norton.
        pages cm
    Includes bibliographical references and index.
    ISBN 978-0-226-19731-9 (cloth : alkaline paper) — ISBN 978-0-226-19745-6 (paperback : alkaline paper) — ISBN 978-0-226-19759-3 (e-book) 1. Sustainability—Philosophy. 2. Sustainable development—Philosophy. 3. Environmental policy—Philosophy. 4. Sustainable development—Decision making. I. Title.
    GE40.N67 2015
    338.9′27—dc23

                                                                                2015012606

♾ This paper meets the requirements of ANSI/NISO
Z39.48-1992 (Permanence of Paper).

In memory of David Fate Norton, who led the way
1937–2014

CONTENTS

ILLUSTRATIONS

Figures

Tables

Boxes

PREFACE

We talk more and more—on campuses, in the press, at international meetings, in commissions, and in the political arena—about sustainability; but do we really know what we are talking about?

One often hears references to "sustainability science," as if the growth of scientific knowledge can propel society toward sustainability. This faith in science, however, can distract us from wondering which values should guide the search for sustainability. As scientific discourse has come to dominate the conversation, the need for better understanding of the important role of values and evaluation is obscured.

This book starts from, and responds to, an important disconnect between the public discourse on sustainability and the academic discourse it has provoked. Politicians use the term "sustainable" with insouciance, as if listeners know what is meant by it. But beneath this thin layer of conventional confidence lies an intellectual brouhaha. Economists argue that sustainability is a term definable within mainstream economic growth theory. Ecologists insist that sustainability requires "resilience" of ecosystems, which can only be measured by ecologists. Meanwhile, environmental ethicists insist that sustainability must rest on normative or moral principles beyond normal obligations to other humans, yet such considerations—however important—have left hardly a mark on either the political or the technical arguments dominating the discourse.

Can these competing fields be amalgamated into an overarching theory of environmental values that gives adequate attention to the theoretical and methodological particularities of all these disciplines? Not likely. Attempts have so far proven impossible because the competing disciplinary theories assert contradictory assumptions and define key terms in incommensurable ways. Will this be a temporary lacuna? Will progress in science unify theory? In fact, there are powerful arguments that no model can capture all the behavior of a complex dynamic system, and environmental problems (especially including the specification of definitive conditions for sustainable living) defy solution by computational models.

I will show that most environmental problems are "wicked problems," meaning they have no definitive formulation and no definitive solution (chapter 2). Wicked problems thus cannot be adequately addressed by scientific or technical means because problems are wicked to the extent that the differing values of deciders set them at odds regarding the nature of the problem itself and frustrate attempts to reduce problems to ones of technical means. This pessimistic conclusion is discouraging both to theoreticians and to those who see the goal as one of identifying—and then achieving—a specific goal that represents sustainability. But what if the theoreticians are asking the wrong questions? And what if these questions have diverted public attention from alternative ways of pursuing sustainability?

The strategy pursued here will, one might say, start at the "other end of the problem-solution pipeline." Economists, philosophers, and ecologists argue about what in nature is valuable and why, attempting to define one or a few goals that would be, according to their specific theory, "sustainable." These diverse conceptions of future goal states, developed within differing disciplines, reflect conflicts over the "nature" of environmental values. The logic of the general multidisciplinary debate about the nature of sustainability, when started from the theoretical end of the argument, has led to polarization and conflict as each discipline attempts to characterize sustainability by using their disciplinary concepts. The disconnect mentioned above—between the enthusiasm for the concept in international politics and the behind-the-scenes turf war over who will own the term and what it will mean—results from clashes over theoretical origins. Often such stalemates require innovative battle plans. Rather than begin with a theory of environmental value that will lock in certain objectives to be pursued in each situation, I will begin with problems and the decision situations in which problems are addressed. In short, I will attempt to describe a process that could lead to an incremental learning approach to achieving sustainability. It is important to note the use of "could" here—suggesting that there is a way forward toward sustainable policies but suggesting also that the opportunity can be missed if the process is not efficacious. My goal is to develop a procedural account of how, if the conditions are right and the correct steps are taken, a community can learn its way toward a more sustainable future.[1]

This goal implies both a negative and a positive aspect, which correspond to the two parts of the book. First, it is necessary to undermine the widespread assumption that we can, in principle, use sophisticated disciplinary models and tools to define goals of sustainability in advance of actions taken to achieve those goals. Given the pluralistic values of contemporary societies, as well as the difficulties in expressing those values in a single disciplinary vocabulary, there cannot be a single, bottom-line definition or measure of sustainability. Disciplinarians, full of faith in their methods, proceed as if more research and more modeling will result in defining an "optimal" state of "sustainability." They reveal their underlying assumption that, in principle, their disciplinary methods will allow a definition of sustainability that is consistent with their home discipline and makes use of its theoretical structures. By undermining these assumptions, I hope to clear away the confusing underbrush, allowing a view of a simpler landscape in which to break a path forward.

Having exposed the goal of a theoretically ideal, disciplinary definition applicable to all sustainable communities as a myth, the second, more positive, aspect of the argument will be to show that, by paying careful attention to processes of decision making, it is possible to sketch a learn-by-doing program with which, under the right circumstances, communities can learn their way toward sustainable living. Readers of this book are thus humbly asked, as they read and speculate about the possibilities explored here, to indulge the author with a temporary cessation of cynicism and to speculate about what *could* happen—what could happen, that is, if environmental policy were consistently approached more adaptively and less ideologically. It is easy to be skeptical about the efficacy of public processes, and in chapter 9 I will examine some issues that can derail collaboration.

More generally, readers may justifiably be concerned that my approach here is mainly oblivious to the pervasive effects of political power and the associated pressures that emanate from formal organizations, ranging from bureaucracies to electoral politics. Here I focus on a different question: what can happen if a community engages in an informal consideration of possible paths forward into the future? This question assumes a political context in which institutions are sufficiently responsive to allow an emergent process of deliberation, experimentation, and communal action in the service of social learning. I do not mean to suggest that power relations and existing governmental institutions are unimportant. I have, however, divided into two steps the challenge of participative governmental action in the search for sustainability. The first step, that of ensuring that the political system is sufficiently responsive to grassroots initiatives to make room for deliberative processes and social learning, involves creating the conditions under which concerned citizens, by following effective processes and practices, have a chance for success.

This book does not address this first step, leaving as a separate task one of learning how, in a society in which special interests and power centers dominate, those conditions can be created and nurtured. For example, the US state of North Carolina has failed to correct, or even to seriously address, failing coal ash ponds created by coal-fired power plants because Duke Energy has captured the regulatory process and blocked effective regulation of its activities. In such situations, an adaptive, collaborative process is disabled. I explicitly skip over this first step, leaving as a separate task to another author to advise how, in a society in which special interests and power centers dominate, those conditions can be created and nurtured. Addressing these topics would require a most astute political scientist; the result would be a very different book than this one. Here, I focus on how a deliberative process might lead to social learning, adaptive management, and improved policies in situations where the necessary conditions for such progress exist.

While failures have occurred in many collaborative attempts at management, the approach here will be to argue that, no matter how many failures can be found, there are also successes, situations in which groups form and develop successful processes that allow cooperative decision making (Wondolleck and Yaffee, 2000; Keough and Blahna, 2006; McShane et al., 2011). Skipping over the first step, then, can be justified by recognizing that, while some situations seem destined for political gridlock, in many situations progress has been made. Consider, for example, the important shifts in understanding and politics that took place in the Chesapeake Bay region, as agencies and the public responded aggressively to problems of pollution of the bay (section 9.4) once scientific study had determined that it had been overnutrified as a result of agricultural and residential runoffs.

What matters in these situations is process: some processes lead to cooperation and better outcomes, while others end in bickering and deadlock (Wondolleck and Yaffee, 2000; Walker and Hurley, 2004). It would require another book to carefully articulate the background conditions that enable cooperative action in response to environmental problems.[2] This book, however, focuses on those efforts that have succeeded in building processes for community decision making and cooperative management. The goal is to sketch how, given that participants bargain in good faith and commit themselves to finding cooperative solutions, communities can learn their way toward sustainability. Better processes—however we eventually define them, the task of this book—are more likely to lead to better decisions than are bad processes. This admittedly is a working hypothesis; I will contrast it with, and test it against, competing, more substantive approaches to evaluating and guiding policy, such as the hypothesis that doing a detailed cost-benefit analysis (CBA) will be sufficient to identify a rationally justified policy.

Central to our proceduralist argument will be a commitment to what I call "normative sustainability," which treats sustainability as primarily a value concept: we must know what is truly valuable to us if we are to know what we should "sustain." Learning about sustainability is as much learning what is valuable as it is learning from science. Given this commitment—to be developed and defended throughout the text—I am committed, in turn, to provide a means to evaluate environmental change as a necessary antecedent to action. So this book offers, as an inevitable demand of my normative theory of sustainability, an alternative to what are today called "economic valuation" and "environmental ethics." That is, I must sketch a procedure by which a community, committed to learning to live sustainably, can progressively rethink and revise many of its current values and policies. I believe that progress toward sustainability is currently hampered by the lack of a neutral field of study of environmental values, which today seems impossible because of the theoretical polarization of the discourse on environmental values. I would be greatly pleased if, someday, historians described this book as the "first textbook in the newly flourishing field of environmental evaluation studies."

ACKNOWLEDGMENTS

This project, which has been under way for half a decade, was enabled and assisted by many friends and colleagues who have contributed to my thinking about sustainability, either by reading and commenting on earlier drafts or by contributing important ideas in informal interactions. These include Barry Bozeman, Marilyn Brown, Dennis Bowker, Rachelle Gould, Lance Gunderson, Paul Hirsch, Eugene Lewis, Steve Minta, Ben Minteer, Douglas Noonan, Anne Steinemann, Marc Weissburg, Hal Wright, and Asim Zia. I have also benefitted enormously from interactions with my students in classes in environmental ethics and in advanced environmental policy, who struggled through earlier drafts and provided invaluable input.

My participation in three invited workshop-style seminars on the topic of ecosystem services brought me up to speed on this increasingly important concept and enabled me to assess when this concept will contribute to—and sometimes harm—public discourse on environmental policy. These included a three-year workshop on cultural ecosystem services organized by Kai Chan and Anne Guerry at the National Center for Ecological Analysis and Synthesis at the University of California, Santa Barbara; a workshop on the ethics of ecosystem service protection efforts, held on the beautiful island of Vilm, Germany, and organized by Kurt Jax and Konrad Ott; and a workshop on cultural ecosystem services organized, again, by Kai Chan and his colleagues and students and sponsored by the Peter Wall Institute for Advanced Study at the University of British Columbia.

The book has also been improved by my correspondence with anonymous reviewers assigned by publishers and, especially, through the comments and challenges from Christie Henry and Amy Krynak in the editorial offices of the University of Chicago Press. Johann Weber assisted in gathering references. David Fate Norton advised me with regard to the publication process. Raquel Asturias, RaGraphics LLC, created the characters Optim and Adapt and the cartoons in which they appear, while Rafael Meza digitized the other figures and tables and helped with all the steps in taking the manuscript to its final form.

PART 1

Change, Complexity, and Decision Contexts

CHAPTER 1

Responding to Change

1.1. Waves of Change

Humans crave stability and security, but everywhere they encounter change. And perhaps nowhere is there more dynamism than in the environmental arena. Rapidly changing systems and problems pervade, which also change in the way they are perceived and in the demands they make on citizens. In 1989, the sociologist Robert Mitchell noted that two "waves" of environmental problems had swept through the first six or seven decades of the twentieth century. He pointed out that original problems of conservation and the environment, the first-generation problems, usually involved threats to a single species or concerns about pollution from an identifiable source. By the 1960s, however, it had become clear that a new wave of problems challenged society, ones that spill over local boundaries and affect entire regions. Mitchell described these problems, including air and water pollution affecting larger systems, proliferation of chemical wastes, and widespread use of pesticides, as the second generation of environmental problems (Mitchell, 1989). He described these new problems as more pervasive, the effects of which were often delayed and subtle yet still capable of creating serious problems in much larger systems and requiring at times the modification of whole industries. Conservationists and environmental groups had to scale up their models and efforts to address these second-generation problems.

Today's emergent problems, however, exist on a global scale, with current policies requiring further expansion in the face of worldwide biodiversity loss and anthropogenic climate change. The rise of global climate change as a dominant issue in contemporary environmental politics will again require the rethinking of national, regional, and local environmental policies, since projections of climate change reveal threats to almost every conservation goal. To continue Mitchell's progression, we now, breathlessly, face third-generation problems that manifest at the global scale even as they portend huge negative impacts on all smaller scales. Addressing the progressive generations of environmental problems is distressing because, unlike with natural populations, the older generations of problems have not disappeared. Complexification results.

Third-generation problems accost us even as we are still off balance from the shift from first- to second-generation problems, as controversies over management of resources continue between federal and state governments and between governments and landowners. Meanwhile, the third wave of global impacts raises a host of new problems, ranging from our lack of knowledge about the functioning of global biogeophysical systems such as the atmosphere to the problem that we have weak or nonexistent global institutions and organizations.

While the shift to global thinking requires a radical leap into the virtually unknown, we are not without guidance as well as useful models for expanding the scale of both problems and processes to address them. Aldo Leopold died before third-generation problems were on the agenda, but he provided a useful conceptual tool—a powerful simile—that can guide scientists, managers, and policy advocates into the larger-scaled problems that emerge at the global level.

Before pursuing Leopold's approach, I note that concurrent with the expanding scale of environmental problems, the public discussion of environmental problems has come to be framed, in policy and in politics, in terms of "sustainability." This concept has, as both its greatest strength and its greatest weakness, broad appeal in the political arena. As a result, advocates for quite different policies embrace the concept, and it has become a "contested concept," as practitioners in multiple fields and from multiple perspectives vie to stake a claim to its definition (Jacobs, 1999).

Sustainability as a concept is versatile, as it can be applied at many different scales. For example, an individual's use of water or energy can be "unsustainable"; so too can the development patterns of a sprawling urban area. One policy, compared to another, might be more sustainable. As climate change invades the environmental agenda, it is necessary to develop a concept of sustainable use of the earth's atmosphere, and ideally that concept should encompass the multiple scales at which humans struggle for survival and longevity, personally and culturally.

For some, this apparently undisciplined use of the term justifies ignorance or myopia. I take a more action-oriented strategy. While noting the myriad scalar applications of the sustainability concept, I try to develop a theory of spatiotemporal relations that allows for a multilayered conception of environmental challenges that embeds individual, local, regional, national, and global scales into a particular—and comprehensive—approach to sustainability.

This effort requires navigation over multiple scales in a dynamic system; the way is made more difficult because the systems in question mix natural and social forces, requiring models that help us understand human adaptation within these interrelated dynamics. I will soon turn to the wisdom of Aldo Leopold, wisdom that was earned by mistakes as well as successes and, using his idea of multiscalar modeling of problems, I begin to develop several concepts that will promote the beginnings of cross-scale integration in rapidly changing environments. The key to that integration will be a recognition of both change, which happens at all scales but at different paces, and the constant uncertainties that come with facing many situations for the first time. Leopold encountered the importance of scale when, in an attempt to increase a huntable deer herd, he caused unwanted changes on the scale of the "whole mountain" (ecosystem). So Leopold's understanding of scale will build a bridge to a process called "adaptive management," an approach that consciously addresses change, uncertainty, and the necessity of learning one's way toward sustainability.

The themes of sustainability and change, interwoven throughout this book, will necessarily require an improved understanding of how to evaluate change over time, especially human-caused change. The image of "ripples in a pond" is often used to illustrate how effects reverberate through time and space. While this metaphor has become a cliché, its overuse is understandable because it can help to simplify the bewildering world of intertwining causal chains that trace the effects of one's actions outward in the world. The emphasis here is on "simplify." In fact, all actions within the complex, dynamic systems that form human environments have impacts that go every which way and make varying impacts on a multitude of natural and social dynamics, from atoms to ecosystems, even as global change makes and remakes those systems. Somehow, within the maelstrom of accelerating change, setting sustainability goals and policies must offer a way of separating good or neutral change from catastrophic outcomes or from even the gradual deterioration of environmental systems.

What is required to become sustainable? To begin to answer this question, I must distinguish between anticipated impacts that threaten the values of a community and those that support such values. In other words, what is needed is a method for evaluating change, a method that—unlike the current evaluative methods available today—can lead those with conflicting values to converge on actions that are agreed to improve unacceptable situations. This reasoning justifies my insistence that an adequate understanding of sustainability will require a new approach to evaluating environmental changes.

No doubt codeveloping a concept of sustainability, together with a more effective means to evaluate changes, seems a daunting task, given the pace of change, the complexity of effects of the actions of technological humans, the lack of basic knowledge of systems within which humans act, and now a recognition that, within this complex dynamic, it is necessary to pay attention to many scales of change and multiple dynamics in natural and human systems. Is it reasonable to think that humans can actually evaluate proposed actions and policies, and then decide whether the impacts of these actions and policies will be harmful or helpful—not only immediately, but also on multiple scales of future time? Could we humans possibly have enough knowledge and foresight to address such complex questions? Do we have any choice but to try?

What if these are not even the right questions to pose? One might expect that the fields of environmental economics, ecological economics, and environmental ethics would provide effective tools for evaluating environmental change; but they, too, are asking the wrong questions. These fields, rather than contributing to a consensus as to which environmental actions and policies should be undertaken in specific situations, have engaged in a polarized war over theories and even words.

Unfortunately, the dominant theories in these fields define environmental values in narrow and theoretically incommensurable ways, and the result has been severe polarization between philosophers and economists, as well as an associated bifurcation of policy discourse into two incompatible, untranslatable vernaculars. As a result, no field of study seems capable of providing a synoptic approach to the task of evaluating ecological and environmental change.[1]

The problem at hand can be stated quite simply: we must find how to provide a method by which to "evaluate" changes in the environment, one that will turn observations of measurable changes in a physical system into either positive or negative evaluations of that change. Speaking intuitively, the goal is to characterize the human value afforded by a physical system at one time, then to characterize that total value again after an actual or proposed change, and finally to compare those total values. To proceed with this commonsense approach, however, it is necessary to agree on which values will count and which methods will be used to define and measure them. Unfortunately, these basic questions of the nature and measurement of environmental values remain controversial; answers to them are too contentious to provide a sound basis for a comprehensive approach to evaluating environmental change.

Consider, as an example, the decision faced by the US Fish and Wildlife Service (FWS) when it undertook to reintroduce wolves into Yellowstone National Park by engaging other federal agencies and national nongovernmental organizations (NGOs) in a decade-long effort. This case represents only one example of the difficulty of maintaining populations of top predators throughout the world. Wolves had been exterminated in Yellowstone in the 1870s, and the ecological system had evolved in new directions subsequently. An unnaturally large herd of elk—highly popular with tourists yet damaging to vegetation and watercourses—had developed. FWS, in the late 1980s and early 1990s, faced a dilemma. It had a mandate, based on the Endangered Species Act of 1973 (ESA), to reestablish breeding populations of wolves in critical habitat from which they had been extirpated. But it had to convince other government agencies at the federal and state level, as well as politicians and their constituents, of the need to translocate wolves from Canada to Yellowstone. This decision was particularly difficult because members of local communities—mostly ranchers and those who support ranching communities—strongly opposed the reintroduction, while far more citizens and environmentalists (many living elsewhere and less affected by wolves if they were introduced) advocated for reintroduction.

What if the FWS had set out to comprehensively measure the values that would be affected by a reintroduction? First, it would have to weigh the costs to ranchers from predictable wolf kills of their livestock and somehow balance those losses against the gains to tourists and tourist-related businesses that predictably increased visitation to the park. Ecologists had insisted for decades that the Yellowstone elk population was too large—an artificial condition at least partially related to the extirpation of wolves over a century earlier—and that it was causing ecological damage. One might, for example, try to measure the preference of "affected parties," but how would one balance the interests of a tourist from the East Coast in seeing wolves against the strong wishes of ranchers to protect their livestock? Even more difficult, how would one measure and take into account the ecological changes that would take place, over generations, if wolves were returned?

Despite the huge controversy, wolves were, in fact, translocated from Canada and reintroduced to Yellowstone, beginning in 1995. What has happened to the ecophysical system subsequently has surprised many people, including some ecologists: "Yellowstone researchers William Ripple, Robert Beschta, and others suggest that since the reintroduction of wolves in the park in the mid-1990s, cascading effects have included a reduction in the elk population, changes in herbivory patterns, a reduction of mesopredators (e.g., coyotes), and recovery of woody browse species" (Eisenberg, 2011, p. 14).

This case illustrates just how difficult it is to know what impacts will occur, let alone to assign monetary or any other measure to the values involved. For example, the effect of coyote population reductions on raptor species was a surprise to many. And, even if it had been predicted, how would one measure (in advance) the value of increased raptors? Similarly, it was only after the experimental reintroduction of wolves that it was possible to study changes in aspen tree populations; yet even if those impacts had been predicted, how would one put a dollar value or any other measure on those impacts? How would one compare, for example, the losses to ranchers from wolf predation against the gains in aspen recruitment and in eagle populations? Environmental values present themselves at different scales of both the social and the ecological system. Wolves affect the individual economic situation of ranchers and the situation of motel owners, one negatively, one positively, but after the reintroduction, unforeseen effects occurred on scales often not expected to be affected by such a decision. Examples like this suggest that humility is appropriate when it is advised that decisions affecting many individuals, groups, and ecological scales should be made by toting up all the effects of possible actions.

One would think that, given the importance of evaluating change as one aspect of understanding sustainability, practitioners of fields that study environmental values would have evolved a unified means to evaluate possible changes to environmental systems. Unfortunately, however, economists and environmental ethicists have found themselves in a theoretical stalemate. Economists, suggesting that we use cost-benefit analysis (CBA), propose using economic measures to estimate the aggregated economic values embodied in the system before an intervention, and then compare these total values with an estimate of the aggregated value of parts and aspects of the system following the mandated change. Environmental ethicists, by contrast, while not questioning the particular assessments of value offered by the economists, argue that the latter fail to understand the "inherent" or "intrinsic" values of nature. Ethicists propose theories about how to understand, identify, and respect this special kind of value, and while they cannot rival the precision of economic techniques, they provide instead arguments that economists' entire conception of value is reductionistic, unreliable, and incomplete. The debate cannot get down to specifics about how much value to assign to specific parts because it stalls on the prior question of the nature and measurement of value.

I do not expect to resolve this deep disagreement about theory, as it may not be resoluble. My goal, however, is to argue that the disciplinary contestants conceptualize the problem inadequately and that, if we ask different questions, progress can then be made *around*, not *through*, this controversy.

Suppose we accept the empirical fact that, in modern, diverse societies, people value nature in multiple ways, and suppose that, instead of measuring identified values, there is a shift to a more process-oriented view by which one evaluates not outcomes but actionable decisions. In this approach, the goal is to seek a guide for policy that recognizes the plurality and also the dynamism of values, and to seek a process by which values can evolve. The key to a more coherent discussion of sustainability is to criticize and replace monistic systems of value—approaches to evaluating change that assume all values can be represented in a single measure—with pluralistic approaches that try to achieve a reasonable balance among multiple values not reducible to a single metric.

Rather than reduce all human values to one kind of value to allow comparison across times, pluralists accept as a fact of life that multiple values will be expressed in environmental policy discourse. From this pluralist standpoint, they can set out to find a fair, effective process that will protect as many of these plural values as possible.

An important consonance of this approach is that, by adopting and celebrating both diversity and pluralism, it becomes easier to conceptualize values as they emerge at different scales of space and time. A pluralist, for example, can recognize that physical systems incorporate multiple dynamics that unfold on multiple scales and, as they unfold, they create and support processes that, in turn, support multiple human values. The pluralist, therefore, can admit that wolf reintroductions, while apparently highly effective at rebalancing disturbed large-scale ecosystems and at protecting the biodiversity of the entire system by regulating the populations of grazing animals, also register negative values at the scale of privately owned and managed ranches in the area. Economists have not succeeded in placing a dollar value on the increased integrity of the system and, even if they could, one hardly knows how to aggregate the value of ecosystem integrity together with the negative value to cattle ranchers. Ranchers balance their books quarterly or annually; conservation ecologists balance their books across decades, even centuries. What is needed is not more detailed knowledge about ecological impacts (though all knowledge in this difficult area is welcome), but rather a system of decision making that is capable of striking a fair balance across these values that emerge on different scales of time and are supported by different ecological processes. Once the incommensurability of cross-scale value conflicts in Yellowstone National Park was recognized, the society-wide value of biodiversity was protected when a private NGO stepped in to provide a fund to reimburse ranchers losing livestock to wolves.

The path toward sustainability or catastrophe (or some state in between) must involve bringing social values—and perhaps individual and private values as well—to bear on choices that are faced by societies at all scales and in all geographic situations. Thus two sets of problems arise that reflect each other: on the one hand, there is a set of causal forces that will play out on our environment, as human actions and developments, as well as natural forces, affect natural and human-constructed physical systems. On the other hand, human societies are formed of local groups constituting larger alliances, institutions, and associations, ranging from families and clans up to and including nation-states. These societies likewise form a complex, many-scaled system. It is important to pay attention to both these hierarchies of scale—one mainly natural (though typically affected to some degree by human action) and the other a system of human-created, polycentric, and multilayered institutions and associations.

The environmental philosophers O'Neill, Holland, and Light have suggested that conservation issues arise because human and natural histories play out on various temporal scales, as human time—and the pace of cultural change—is so much more rapid than natural, ecological, and evolutionary changes. This mismatch, they suggest, lies at the heart of the problem of how to manage for the future in the present. Addressing this mismatch, seeking to tell a narrative that makes sense of the clash between the human sense of time and the slow workings of natural systems, "is about negotiating the transition from past to future in such a way as to secure the transfer of . . . significance" (Holland and Rawles, 1994; O'Neill et al., 2008, pp. 155–58).

In seeking out better and more appropriate processes of decision making, it is necessary to shift attention to the decision contexts in which many environmental problems are encountered. In many instances, environmental problems and policies proposed to address them are highly controversial and confusing with regard to which values are most important, how to express those values, and how to achieve some kind of cooperative action in intractable situations. These confusions emerge when competing groups with conflicting concerns and interests seek varied outcomes as they contest environmental policies within a complex social context. With a shift in focus away from theories about what is valuable to a discussion of how multiple values play out in day-to-day decisions about the environment, it is easier to see that environmental economists and environmental ethicists have been asking the wrong questions. Even if answered, the answers do not resolve the kind of complex problems actually involved in protecting multiscalar environmental goods and sustainability in practice. What kind of system of evaluation is likely to help communities make difficult decisions in local situations of great controversy? Directing attention to the decision context turns a seemingly impossible theoretical dilemma about the nature of environmental value into concrete activities that can protect environmental goods in actual situations with unique issues. The goal, then, must be to determine which values should be protected in particular situations. This book offers a process by which democratic societies can make such determinations.

1.2. Strategies for Achieving Sustainability

There are two main strategies by which one might explore sustainability goals. The first is an optimizing strategy, and the other is an "incremental learning" or adaptive strategy. To more sharply contrast these two strategies, I have created two characters—I'll call them "Optim" and "Adapt"—to suggest practices appropriate to their favored strategy, which is illustrated in figures 1.1, 1.2, and 1.3.

The optimizing strategy identifies a small set of characteristics or a formula for identifying systems that are "sustainable." By navigating toward that objective, one could, in principle, identify the "most sustainable" of possible outcomes by comparing them to the optimal outcome. Accomplishments can then be judged by measuring progress toward the stated objectives that correspond with a substantive ideal of sustainability.

By contrast, an incremental learning (or adaptive) strategy starts with the assumption that one cannot, in advance, specify with any accuracy the objectives that would constitute having achieved sustainability. In this strategy, the emphasis shifts to understanding and specifying a process by which a community or society could, by following certain promising steps and applying insights from the social sciences, learn its way toward sustainability. This assumes that communities can learn and grow with respect to their values, modifying their goals en route, as well as use scientific and technical means to improve their understanding of system behavior. The incremental learning strategy also reacts to problems with experiments and actions. Since this does not specify objectives or goals of sustainability at the outset of inquiry, these must be constructed on the fly, even as actions are undertaken. This distinction is illustrated in figures 1.2 and 1.3.

Figure 1.1. The road to sustainability: Optim and Adapt offer their services as guides to achieve sustainability.

Figure 1.2. Two possible guides: Optim and Adapt offer different types of advice. Optim calculates based on theory, whereas Adapt guides a process that will encourage learning by doing.

OPTIM ON SUSTAINABILITY

1. Define objectives that would describe success in achieving sustainable living.

2. Develop methods for calculating optimal success in achieving objectives.

3. Develop measures that indicate degrees of success in achieving objectives.

4. Provide a road map for the sciences to determine and prescribe means to achieve the objectives.

5. Success of policies and actions will be judged for their contribution to achieving the stated objectives.

6. If achieving optimal success is not achieved, alternative outcomes should be ranked according to their approximation to the optimum.

Figure 1.3. Optim on sustainability: Optim tries to articulate outcomes and objectives that will maximize values according to a theory.

An analogy from evolutionary thought illustrates these contrasting strategies. Charles Darwin, in *The Origin of the Species*, wrote of "artificial selection" and explained that one of his most potent clues in his search for a physical model of evolution was the remarkable success of plant and animal breeders, who over millennia have developed techniques to selectively breed

ADAPT ON SUSTAINABILITY

1. Recognize that both sustainability goals and means to them are fraught with uncertainty. Resolve to apply The Scientific Method to all controversies when possible.

2. Pay attention to process and try to create a situation in which social learning —both about goals and means—can occur.

3. Develop concurrently appropriate institutions and a public discourse capable of recommending actions and building trust.

4. Explicitly discuss the choice of important variables as indicators of progress toward sustainability.

5. Engage in an ongoing examination and re-examination of proposed policies and goals even as the nature of problems shifts.

Figure 1.4. Adapt on sustainability: Adapt recommends developing an iterative process based on community deliberation and learning by doing.

for specific traits in plants and animals: "We are profoundly ignorant of the causes producing slight and unimportant variations; and we are immediately made conscious of this by reflecting on the differences in the breeds of our domesticated animals in different countries—more especially in the less civilized countries where there has been but little artificial selection" (Darwin, 1859–64, pp. 197–98).

Plant and animal breeders have for millennia created distinctive breeds and lineages, in some cases creating very cute or very ugly breeds, and in other cases improving by increments one or a few characteristics of special importance to breeders or their customers. The art of artificial breeding is slow and painstaking, requiring selection over many generations, but persistent, multigenerational selection demonstrates beyond any doubt that phenotypes can be manipulated, within certain bounds. They can also be carried to the extremes, as in the case of Great Danes and Chihuahuas (see figure 1.5). In the latter case, the goal is to create highly specialized breeds with distinct characteristics that appeal to narrow interests; yet selective breeding can also endeavor to produce less-specialized, robust animals with multiple strengths and versatile behaviors.

The process of artificial selection provides an analogy that will clarify the two strategies I am contrasting. Pretend for a bit that you are a technically skilled dog breeder and that you live in a time or a place where most dogs are fairly close in both genotype and phenotype to wolves. Now, to tell my story, I need you to imagine that you live for many generations or that your children and grandchildren will continue your efforts, so the evolutionary pattern becomes visible—one that is hard to see in the short run because the selection manifests itself in tiny increments over multiple generations.

Recognizing that you are technically good at realizing a genotypically based phenotypic change, you face a choice regarding the strategy by which you will exercise your skill. Will you decide to go for highly distinctive breeds, ones that can do special tasks or appeal to rich people who want a dog that will win "best of breed" in a dog show? The alternative is to stay closer to the original wolflike stock and incrementally adjust the breed with respect to characteristics appropriate to diverse tasks and roles. The first alternative, then, is analogous to the optimizing strategy mentioned above; it aims to maximize one variable, such as size, fetching ability, and the like, or perhaps it selects for several characteristics, such as strength and aggressiveness. An alternative strategy, favoring more robust animals, would pay attention to a much larger number of variables and would seek, through an adaptational strategy, to incrementally improve some characteristics of the breed, keeping in mind that more general fitness for varying environments and multiple

Figure 1.5. Great Dane and Chihuahua: Artificial breeders trying for specialized breeds have developed dogs with spectacular characteristics. Photo by Finch, Wikipedia (http://en.wikipedia.org/wiki/Image:IMG013biglittledogFX_wb.jpg).

roles may require specific characteristics that will be achieved while trying to maintain basic, core skills that will prove useful in many situations.

The first strategy, aiming at distinctness in particular characteristics, attempts to maximize certain characteristics such as large or small size and attempts to move the phenotype of the breed in ever more distinctive directions. The result of this optimizing strategy, which can lead to success if a breeder can develop a lineage that exemplifies to an extreme degree the distinctive characteristics of a special breed, will appeal to a limited, specialized market but will not result in a dog that is versatile in diverse and changing situations.

The second, adaptational strategy, on the other hand, depends on maintaining basic fitness that will allow an animal to function in a range of circumstances and roles. Here, a fairly long list of characteristics of the animals will be perpetuated—such as intelligence, long life span, freedom from disease—even as adaptational breeders seek to incrementally improve some particular characteristics such as ability to fetch shot game or an especially sharp sense of smell for tracking.

In the case of animal breeding, neither one nor the other of these strategies is the "correct" one. Both have a role in the artificial selection business, and each strategy will be appropriate in some circumstances and not in others. But this is not to say that there are not important tradeoffs involved in making the decision as to which strategy to follow. It is well known that breeding any species for extremely specific characteristics, such as size or a particular look of cuteness, generally causes deterioration in some other features. For example, in dog breeding, Great Danes tend to live shorter lives than most dogs and suffer from ailments at least partially due to inbreeding and maximization of a few characteristics that make the breed distinctive. It is these characteristics, plus training, that wins dog shows.

Working dogs, by contrast, must be adaptable to many different challenges. So, following the incremental strategy is more appropriate if one hopes to breed dogs that will be generally healthy, adaptable to multiple situations, and appealing to a wide range of dog users and dog lovers. Especially, dogs trained to support police activities or dogs bred to engage in special operations must be able to do many different things, which requires strength, intelligence, and trainability.

For this reason, dogs bred, chosen, and trained for these purposes are usually German shepherds or Belgian Malinois, because specimens of these breeds have the best combination of intelligence, good sense of smell, speed, strength, courage, and other characteristics necessary to execute a variety of tasks. Bred as guard dogs and herding dogs, they are highly adaptable, and their strengths are related to those of the original wolf stock, since pack hunters of large herbivores require a wide range of strengths and skills. These characteristics, of course, would be useful in a broad range of situations, and the animals produced will be adapted to many situations. I will say that such animals are "robustly adapted"—distinctive in a few ways, perhaps, but also having those qualities that make them adaptable to diverse situations requiring multiple skills.

I hope this analogy helps clarify the two strategies necessary to explain my choice of an incremental learning strategy over an optimization strategy in searching for sustainability. Optim is like the first breeder, who specifies one or a few particular characteristics that she can identify in advance and that she wants to maximize. This strategy has the advantage that, once the goals and objectives are set in advance, pursuit of the optimum becomes a matter of developing effective means to the desired end, while emphasis on management is on means-oriented technical knowledge in the hands of experts. Her animals will thus be altered to accentuate certain characteristics and to eliminate traits not typical of the breed sought.

Adapt, on the other hand, is more like the breeder of less-specialized dogs who seeks to improve his stock incrementally, seeking to develop a breed that has multiple strengths and is more "robust" in the sense that its multiple talents prepare it to be useful in many and emerging situations. Adapt, accordingly, is looking for a more robust conception of sustainability, one that serves a range of values and supports varied options. I can also highlight the difference between the two characters by reference to a historical distinction, suggested by the poet and philosopher Archilochus, who concisely summarized two intellectual approaches, representing one as a hedgehog (here, Optim) and the other as a fox (see box 1.1).[2]

Box 1.1. Archilochus: The Hedgehog and the Fox

The poet and philosopher Archilochus (*ca.* 680–645 BCE) wrote, "The fox knows many things, but the hedgehog knows one big thing." The philosopher Isaiah Berlin picked up on the phrase as the starting point for a clever, playful essay in which he divided intellectuals, especially philosophers, into two broad categories: foxes tend to take problems as they come, remain flexible in response to new evidence, and are wary of broad theories; hedgehogs, on the other hand, interpret experience by embracing a broad theory that explains everything. Foxes are often criticized for believing unconnected facts while missing the big picture; hedgehogs are berated as "ideologists."

As we move through this book, we will be accompanied by two characters who offer to be our guides on the way to sustainability: Optim, a hedgehog, approaches the environment and its problems with a grand theory about the "value" that is found in or derived from the world, and following that theory, he hopes to guide us toward a specifiable state that is sustainable. Adapt, the fox, takes problems as they come, pays attention to local and small-scale situations, expects surprises on the road to sustainability, and tries to learn more from experience than from grand theory.

The difference in sustainability strategies developed here resembles a distinction proposed by Manuel Arias-Maldonado (2012), who, in his important discussion of sustainability politics, distinguishes "open" from "closed" approaches to sustainability. Closed approaches to sustainability take the content of the concept, as well as the "shape of social, cultural and economic relations with the environment," to be "technologically or ideologically determined, leading to a model removed from public debate." As in the reasoning of Optim, if one assumes that sustainability can be predefined, and if one can articulate its goals in advance of action, the problem of sustainability is transformed into a problem of management—that is, as a matter of finding efficient means to already identified goals. As Arias-Maldonado says, in pursuing closed sustainability, "[t]he outcome takes precedence over the procedure" (Arias-Maldonado, 2012, pp. 108–9).

In open approaches, on the other hand, "Sustainability is instead conceived as a necessary societal goal whose content and conditions are nonetheless not pre-given. There is no single sustainability, but a whole range of different, even simultaneous, possibilities." Arias-Maldonado strongly advocates open approaches analogous to those here ascribed to Adapt, and he sees open sustainability as raising broader political questions not addressed under the closed approach: "The politics of sustainability are open to public deliberation and to the spontaneous development of ideas and practices. Sustainability is thus inherently open and radically democratic, since only a democratic society guarantees the necessary conditions for its achievement" (Arias-Maldonado, 2012, pp. 109–10).

Arias-Maldonado writes from the viewpoint of a European political scientist, so he does not frame his open approach as involving adaptive management, the term for a management style that has developed in the United States, yet it appears that his open approach to sustainability captures virtually all the important themes associated with an adaptive approach in which the goals and aspirations as well as scientific beliefs are open to revision. In section 5.4.4, I will return to Arias-Maldonado's conclusions about the role of deliberation within modern liberal democracies.

Readers who are familiar with the excellent book *Resilience Thinking*, by Brian Walker and David Salt, will recognize that my two strategies, indeed my favoring of the latter, parallel their reasoning. They say, "Current 'best practice' is based on a philosophy of optimizing the delivery of particular products (goods or services). It generally seeks to maximize the production of specified outcomes or products from the system (set of particular products or outcomes) by controlling certain others" (Walker and Salt, 2006, pp. 5–6). They here introduce a key concept in current management thinking—resilience, which they explain as "the capacity of a system to absorb disturbance and still retain its basic function and structure" (Walker and Salt, 2006, p. xiii).

Further, they associate the idea of "sustainability" with a rejection of maximization and optimization: "Though efficiency per se is not the problem, when it is applied to only a narrow range of values and a particular set of interests it sets the system on a trajectory that, due to its complex nature, leads inevitably to unwanted outcomes" (Walker and Salt, 2006, p. 7).

Following parallels with the work of these writers, my strategy is to pursue incremental learning and to treat goals, as well as scientific hypotheses, as open for experimentation. Now, however, one can begin to see that Adapt's attitude toward science, and her pursuit of sustainability goals, differs sharply from those of Optim. Optim follows the "maximization" logic criticized by Walker and Salt. Optim would choose a few important outcome variables and manipulate other variables to maximize those outcomes. Economists, for example, may fall victim to Walker and Salt's criticisms because, relying on "comparative advantage" and "economies of scale," they tend to favor concentration on a few natural products like timber or agricultural products at the expense of overall well-being of the system generating these products. In directing production to these limited variables, such a management plan will not do justice to the complex processes that provide ecological support for the plan. This approach will lead "inevitably to unwanted outcomes." Adapt, on the other hand, seeks a more robust management plan and a more robust adaptation as represented by German shepherds rather than a narrowly targeted plan as represented by the pursuit of Great Danes or Pekinese.

I am inclined to follow the advice of Adapt, because the situation we actually face is undeniably a situation of great uncertainty, rapid change, and perplexing wicked problems. Seekers after sustainability won't need Great Danes or Chihuahuas, so to speak; they'll need robust actors that are adaptable. They should seek a route to sustainability that is functional in dynamic situations, following the incremental learning approach of Adapt, because it is more appropriate to the situation that sustainability seekers actually face.

An ideal guide might be able to foresee the future, yet Adapt doesn't have that ability; Adapt has no more foresight than we do as humans. Adapt can only infer what will happen in the future by access to what we humans currently know, when we apply the best science of our day. Adapt is ideal in another, weaker sense: she can draw on all human knowledge and creativity, literature, and art. In this sense, Adapt is equipped with an unerring ability to apply the scientific method to reduce uncertainties that nag scientists, policy analysts, and managers.

1.3. Sustainability: A Contested Concept

The very idea of sustainability can almost sound paradoxical, especially when it is contemplated against the backdrop of constant and accelerating change. What does it mean to "sustain" the dynamic systems within which we live and act? Ecology emphasizes the dynamic nature of physical systems, while sustainability seems to aim at a level of constancy with respect to at least certain aspects of human and natural systems. Can these systems be frozen with respect to characteristics that people living today wish to sustain? Any conception of sustainability must provide some account of what is to be sustained amid rampant change.

Some writers have tried to put sustainability theorists on the defensive by ridiculing a straw man, which can be referred to as "absurdly strong sustainability." According to that kind of sustainability (which, as far as I know, nobody seriously advocates), our generation owes the future exactly what we inherited from previous generations. But, of course, present generations must use and alter the resource base, so complete freezing of change is impossible, for it would lead to extinction of humans. Humans must alter nature to remain alive and must parent the following generation even as changing natural systems affect all evolution, including human evolution, since each generation starts from an altered baseline. A variant on absurdly strong sustainability might consider all the many generations of individuals yet to be born, and it might seem that the present generation has rights to at best a tiny sliver of earth's resources. These concepts of sustainability seem to be strongly biased toward the future, endorsing forms of what one might call "futurism"—the favoring of the future against the interests of present people.

It has been argued, however, that history tells us that people of the future will predictably be wealthier than the present generation. If so, why should one limit consumption to improve their lot? This question might lead to an alternative bias, "presentism," which disregards or disavows impacts on future people. Following this line of reasoning, some have suggested that achieving sustainability requires no more than economic business-as-usual. For symmetry, this position can be called "absurdly weak sustainability." Sustainability must place some constraints on the normally short-term thinking of economics, or there would be no point in having the concept (Beckerman, 1994; Beckerman, 1995; Daly, 1995).

Surely, however, reasonable positions may be taken between these extreme viewpoints; a more nuanced view is essential, which requires posing difficult questions. For example, since some resources are finite and exhaustible, while others are renewable, is it fair to the future if this generation exhausts nonrenewables, leaving them only with renewable sources of energy? Is it a reasonable and fair substitution if systems passed on to the next generation represent sustainable pathways forward, even though the future will face a changed resource landscape? In the background of these debates also looms a momentous question of the role of technology in the effort to be fair to the future. Is technology the problem? Or is it the solution? Is technology a set of tools humans control, or is it at this point in history "autonomous" and impervious to rational critique? These and other puzzling questions will affect any concept of sustainability that allows change but will also hold constant some aspects of natural and social systems that can be the basis for a healthy, happy future for our successors.

Sustainability is a contested concept because there is no agreement about what, in the midst of change, should be sustained for the future. Is there a reasonable, defensible position between absurdly strong sustainability (which would freeze economic activity) and absurdly weak sustainability (which eschews regulation and simply pursues economic wealth at any cost to the environment)? Articulating a conception of sustainability between these extremes requires an answer to the question: How do or how should communities decide what is so valuable to them, their economy, their culture, and their spiritual life that it must be protected for the future?

The concept of sustainability—often paired with "development"—gained prominence in the 1970s and, especially, in the 1980s. Many of the issues today gathered under the heading of "sustainable development" were first raised in the United Nations Conference on the Human Environment, sponsored by the UN Environment Programme, and held in Stockholm, Sweden, in 1972. The eye-opening report of the Club of Rome, *Limits to Growth*, was published in the same year, triggering both acclaim from environmentalists and denunciations from progrowth forces (Meadows et al., 1972). The term "sustainable" and the related concept of "sustainable development" were highlighted in the much-discussed report from the World Commission on Environment and Development *Our Common Future* (1987). The report itself is often referred to as the "Brundtland Report," after its chair, Gro Harlem Brundtland. The report defined sustainable development as "development that meets the needs of the present without compromising the ability of future generations to meet their own needs." This definition has stimulated considerable discussion because it characterizes sustainable development mainly in terms of resources available to people. Definitions such as this will, later in this book, be subject to criticism for not explicitly mentioning requirements with respect to the ecological and physical functioning of the systems within which humans live.[3]

Note that various approaches to sustainability and sustainable development have advocates among the practitioners of several disciplines, who connect the concept of sustainability to their own theories, concepts, and models. In particular, the social scientific field of economics and the natural scientific field of ecology have attempted to articulate sustainability principles that embody the knowledge and wisdom of their fields. Indeed, once authors from these two fields had offered their insights on the topic of sustainability, it became obvious that they were offering conceptions that differed in their most basic assumptions and implications (Norton and Toman, 1997). Faced with the apparent conflicts that these insights seemed to imply, a group of interested and creative scholars from those two disciplines—as well as interested parties from different but related fields—founded a new field, called ecological economics, which is sometimes referred to as "the science of sustainability" (Costanza, 1991; Clark and Dickson, 2003).

While the concepts of sustainability and sustainable development are contested, sustainability does have a core meaning: it pertains to relations across generations. However different the disciplinary, conceptual, and political approaches taken by various theorists to understanding and measuring sustainability, one would not be discussing sustainability if one did not in some way focus on our moral obligations beyond the more usual intragenerational moral concerns to the longer time frames of intergenerational morality. Although they often do not explicitly mention the moral aspects of intergenerational relations, choices to define sustainability in terms of economic growth theory or maintenance of resilience imply differing moral commitments.

This projection beyond the present is confusing and contestable because many of our typical moral guidelines don't apply, at least not straightforwardly. For example, within generations, reciprocity can be expected or required as a condition of moral interactions, and one can think of collective moral decisions as reflecting some kind of contractual arrangements, explicit or implicit. Intergenerational morality, however, is inevitably asymmetric in that decisions made today may affect people of the future, though their decisions will not affect the lives of those presently living. While sustainability theorists must recognize this and other apparent discontinuities between intra- and intergenerational fairness, it seems clear that sustainability is about the future and the impacts of today's activities on the future.

Thus the term "sustainable" is not infinitely malleable, despite the many interpretations and contestations among its advocates. The objective, then, must be to develop a conception of obligations to the future that is appropriate, given the moral gravamen of multigenerational concern that most humans feel, while recognizing that sustainability thinking must also account for the unique features of asymmetric moral relations.

For convenience, I speak of the "bequest" that each generation bestows on its successors as the sum total of cultural and physical resources that are passed from that generation forward. Historically, civilizations have often been concerned about their bequest to their future generations but, more often than not, these concerns had to do with maintaining the power and privilege of the noble classes or with maintaining religious traditions more than merely a livable environment. In many historical cases, obviously, the actual bequest was often quite different from the intended bequest, a distinction one must recognize and maintain because, to paraphrase an old saw, "the road to hell is paved with sustainable intentions." Clearly, present intentions are relevant to any goal of protecting the well-being of future people, yet just as clearly, what will truly matter to the future is not intentions but the actual bequest. Since this project is mainly philosophical and ethical, the goal will be to query what would be a "fair" bequest to the future, as well as to clarify the intentions each generation should pursue in order to act sustainably.

I will focus on how a community might go about creating a fair bequest (i.e., provide a specification of reasonable and fair sustainability goals), how best to develop a sociopolitical process that is capable of articulating such a fair bequest for a given community, and how that process can be an integral part of the sustainability project itself. Choices regarding the bequest of any given generation should be structured while one is mindful of the impacts of their actions on the future. Controversy, however, breaks out once one actually tries to describe and measure the contents of a bequest. This controversy has not generally focused on determining concretely which objects and processes should be preserved as elements of a just bequest; rather, the debate is mostly hung up on a cluster of prior, theoretical disagreements about the nature of environmental value and on disagreements about how to measure environmental value. For example, economists generally describe sustainability in terms of economic growth theory, while ecologists insist that sustainable systems must be ecologically resilient.

Although turf wars over key policy conceptions are not unusual, there is an unusual situation with respect to the concept of sustainability: it became an important term politically before it was given a clear meaning within a single discipline. The term, important since the 1970s and 1980s, has been written into political documents and used in international negotiations even as its definition remains contested. I am not able to finally resolve these interminable debates, though I do want to clarify the meaning of sustainability by seeking a process that helps communities learn their way toward sustainability—that is, learning toward a more robust concept of adaptation that pays attention to the broadest range of human values.

1.4. Aldo Leopold and Changing Times

Deeply entwined with the puzzling idea of change and accelerating change is the commonplace notion of time. Yet that commonplace notion, it turns out, is far more complicated and nuanced than it seems. As I think about time and change here, I will follow Aldo Leopold, who puzzled over both of them—and over their reciprocal relationship with humans and human cultures—as he struggled to integrate human activities into the natural scheme of things. While Leopold, as far as I know, never used the term "sustainability," he nevertheless realized that humans must be considered a part of nature. He was also aware that modern humans have developed technologies that, because of their power and pervasiveness, can apparently accelerate change and in some cases can arrest or derail the dynamics of natural systems (Leopold, 1939; 1979; Norton, 2005; see box 1.2).

Box 1.2. Aldo Leopold, Conservationist (1889–1949)

Aldo Leopold graduated in 1909 from Yale University School of Forestry, which had been endowed by Gifford Pinchot's family. On graduation and entry into the US Forest Service, he was assigned to the southwest territories (today's Arizona and New Mexico). The young Leopold encountered a deeply divided conservation movement, because Pinchot, the utilitarian, scientific resource manager, and John Muir, a mountaineer who described the forests as "the people's cathedrals," seemed to point in different directions.

Faced with this conflict, Leopold created a unifying approach that still guides American conservation today. Initially, Leopold followed Pinchot's managerial philosophy: that the purpose of the forests was to support economic development through the wise and fair use of natural resources. As he matured, Leopold's approach to conservation theory and practice changed. He became the first professor of game management at the University of Wisconsin in 1933, where he developed and taught his conservation ideas. He also purchased a burned-out farm on the Wisconsin River and, with his family, put into practice his ideas of management and restoration on the land that now hosts the Aldo Leopold Foundation.

Today, it is generally agreed that Leopold was the greatest conservation thinker in the rich tradition of American conservation, and interpretations of his work and his lessons remain a central topic in the history of conservation. In his authoritative biography of Leopold, Curt Meine said it well: "In part because he held both a Muir-like appreciation of nature and a Pinchot-like intent to use nature wisely, Leopold was destined to lead a life of conflicting desires, constant questioning, and unending effort to better define the meaning of conservation" (Meine, 1991).

Aldo Leopold was a masterful writer; his prose is both sparse and rich, layered with meanings and laced with apt analogies, metaphors, and similes.

A Sand County Almanac and Sketches Here and There, published shortly after his death in 1949, reveals Leopold's wisdom, thread by thread, as multiple ideas are woven across chapters. It tells its story simply, but it is by no means a simple story. One thread of the book is without question the riddle of time itself, time's passing, our confusions about its scale, and the emerging idea that events in our lives are fleeting and ephemeral when compared with the depths of time we humans live within. And yet we struggle, as Leopold says, "for peace in our time," peace that seems always to elude us as times change (1949, p. 133).

One of Leopold's most poetic, and alarming, essays, "Marshland Elegy," is a reflection on how natural systems of great beauty—in this case an ancient crane marsh—can be destroyed by human carelessness and greed (1949). Leopold begins his story standing at the edge of a marsh, which no doubt is a composite of many marshes he had visited; he hears in the distance, in the fog, loud and confusing noises that turn out, after Leopold waits for the source of the sounds, to be an "echelon" of birds that drop through the mist and land, beginning another day on the crane marsh. Leopold's hardly patient wait for the cranes to reveal themselves serves a purpose in his own story: he establishes that humans experience time as duration and are often impatient for things to happen sooner than they will happen in nature. The old saying "A watched pot never boils" is not a comment about pots, but about human perceptual psychology. Leopold's wisdom is best encapsulated as a recognition of the severe disconnects between the time scales at which humans—especially those of us equipped with powerful technologies—change nature and the scales at which change takes place naturally.

To expand consciousness of slower, cyclical changes that characterize natural changes, Leopold introduces another layer of time as he explains the ecological processes by which the cranes and the marsh had coevolved, and then layers on an even longer frame of time: the pace of geological forces that created the great lake in his story and eventually its marsh. In other words, Leopold understood human lives to be embedded in at least three scales of time, three stories that unfold across an array of systems at multiple scales. Human experience of duration is but a glimpse into large, slow-moving processes of ecological and evolutionary change. Leopold's wisdom starts with a recognition of a radical disconnection between the ways we humans understand time and the far slower workings of the natural systems that form the context of our day-to-day lives. That wisdom, he believes, comes from a deep respect for understanding that time-as-duration is not the only time that is important.

Leopold described the cranes as "trumpets in the orchestra of evolution," arguing that their longevity provides them "a paleontological patent of nobility." The crane is the "symbol of our untamable past, of that incredible sweep of millennia which underlies and conditions the daily affairs of birds and men" (Leopold, 1949, p. 96). But he did not leave his reader with simply a placid remembrance of a beautiful natural place. Instead, he tells the story of how the natural place was noticed, damaged, and then ruined by human choices and human greed. The paleontological patent of nobility is, as Leopold writes with bitterness, "revocable only by shotgun."

The reader can marvel at Leopold's tender evocation of life in a less complex time, and can be moved by the trenchant prose he employs to describe the marsh's destruction. For my purposes, however, I can extract from his writings a marvelous tool, a tool for bringing order to our large-scale understanding of dynamic ecological systems and the penchant of us humans to ruin them. Leopold, in the process of explaining the coadaptation of marsh and crane, called attention to three embedded scales of time—human, experiential time (duration); ecological time (the pace at which an ecosystem recovers from a major disturbance like an ice sheet or a fire); and geological time ("deep" time in which humans' natural and evolutionary history emerged). In his essay "Marshland Elegy," the layers of time provide an appreciative accounting of the role of ecological and evolutionary systems in the emergence of the contemporary world in which we live today, most of us ignorant of this incredible story of the entwining of physical and living systems that has created our present world (Leopold, 1949).

Aldo Leopold's layered conception of time, however, is given a sharper edge in a subsequent essay, the only one in which he turns his critical eye on his own actions and policies—the much-discussed essay "Thinking like a Mountain" (1949). Leopold was goaded into writing this essay by one of his students, who insisted that the scientist and ecologist had made mistakes that, based on his current understandings and calculations, were serious, having caused unnecessary harm to the natural systems of the American Southwest (Meine, 1991). By the time Leopold was putting *A Sand County Almanac* in final form, he had learned the importance of wolves in wilderness areas, had founded The Wilderness Society, and was widely regarded as an articulate defender of predator protection in wilderness areas.

He had not always borne that reputation. Leopold became the director of operations for the Forest Service lands in the southwest territory in 1919 and, in an effort to make those lands more economically productive, he initiated a program to eliminate wolves and mountain lions from the territories so as to both enlarge the deer herd and increase hunters' success. Years later, he said, "I thought that because fewer wolves meant more deer, that no wolves would mean hunters' paradise" (Leopold, 1949, p. 130). In 1926, a few years after he completed his wolf-removal program, the predator-free, swollen population of deer was reduced by some 60 percent in a single rough winter, as the huge deer herd ate all available brush and then starved to death. "I was young then, and full of trigger-itch," he remarked, but after seeing an old wolf die, he was forced to conclude that "only the mountain has lived long enough to listen objectively to the howl of a wolf" (p. 129). Leopold learned that getting management decisions right requires that humans consult not just human time—the time scale on which the next season's harvest of deer is anticipated—but that we embed an understanding of our fellow humans' actions and impacts in a more layered and nuanced system in which scales of time can be arrayed in ways meaningful for fully comprehending ecological and evolutionary change.

Leopold used analogies, metaphors, and finally, a powerful simile to nail down the underlying confusions that caused him to err: "In the end," he wrote, "the starved bones of the hoped-for deer herd, dead of its own too much, bleach with the bones of the dead sage, or molder under the high-lined junipers." Using personification, Leopold depicts the destruction of the undergrowth when deer destroyed the "bony" vegetative structure that protects the mountain from erosion, causing the mountain to "fear" the deer for lack of wolves to check the deer populations, especially in wilderness areas where hunters' access is limited or difficult. Eons of soil-building and coevolution of mountain and vegetation were sacrificed for a quick fix to a tourist economy. If humans think only in terms of experiential time, we can "revoke"—by shotgun and other means—paleontological patents of species and the habitats they live within.

Learning to "think like a mountain" is not forgetting how to think like a human actor; rather, it is to see as well the additional layers and complexities of a multiscaled system in which we humans inescapably live. Learning to think like a mountain is learning to see the mountain unfolding on all three of the scales of time mentioned in Leopold's elegy to the marshland. In "Thinking like a Mountain," he uses the three time scales to illustrate his error in deer and wolf management—that the value of wolves was truly manifest on a scale larger than the one he addressed, a scale he had ignored when devising his policy of touristic development. In his original policy, Leopold ignored this larger, ecological scale of species interactions, especially the impact of grazer populations on the vegetative bones of the mountain.

As it turned out, most casual hunters would not pursue deer into roadless wilderness, and deer populations grew exponentially, radiating outward into the entire area, resulting in a population crash. Leopold's simile represents a sort of "Eureka!" moment (however slowly Leopold's actual beliefs and policies changed). What is most interesting about his transformation is that it was holistic: he simultaneously adopted a new and expanded "model" of the situation and also reevaluated the events he was orchestrating. On this new model, the wolves, formerly referred to as "vermin," were now important and valued elements of a complex, dynamic system. Notice—because this will be important in future considerations—that wolves reverse their value as their actions and contributions are traced through the longer scales of time. In one dynamic—the story told by ranchers and hunters—wolves are vermin; in the mountain-sized dynamic, wolves are actually keepers of the integrity of the ecological system.

The transformation in Leopold's thinking, as well as his use of metaphors and similes to explain his insights, can be transformational. The reason that "Thinking like a Mountain" can be taken as a sort of parable, I will argue, is that his transformation represents a change in perspective, and such a change in perspective can involve both a change in factual beliefs and a shift in values. Further, Leopold's use of analogy and metaphor is itself representative of the important changes that must take place in human behavior and policies. The simile of thinking like a mountain is entirely apt because it embodies a transformation in the "model" Leopold used to understand and motivate his actions and simultaneously evokes the human values of stability and continuity.

Taking a cue from Leopold and enlisting a more recent concept that has proven useful in the social sciences, one can say that Leopold became aware that human individuals encounter the world using "mental models" and that these semistable perspectives, which are composed of thought, conceptualization, beliefs, and values, all provide a structure into which individuals, faced with new information, make sense of and integrate that new experience into their ongoing worldview. One can say, informally, that a mental model is the mind-set that allows individuals to function in a changing world that constantly yields new information. Yet individuals do not live in a social vacuum and, as it turns out, when they join groups with similar interests and local perspectives, they tend to develop "cultural models" that allow communication about their common interests. Cultural models, then, are the shared thoughts, conceptualizations, beliefs, and values that support communication among individuals with overlapping mental models.

Leopold incorporated a similar idea into his discussion of his own changing understanding of the wildlife populations and natural systems of the American Southwest. He recognized the extent to which a certain mental model—one widespread in the culture and deeply engrained in the training and thinking of utilitarian resource managers of his day—had narrowed his focus and caused him to miss important factors relevant to the situation at hand. Leopold, guided by a mental model that emphasized short-term productivity, initially eliminated wolves to increase the next season's deer herd, without recognizing the long-term and widespread consequences of freeing a deer herd from the burden of predation.

I believe there is great wisdom stored in Leopold's transformative experiences, in his descriptions of them, and in this wisdom being discussed, expressed as a shift in metaphors for the situation in the Southwest. His simile provides the first clue in understanding sustainability: changing scientific models often involves changing values, and value changes are so mixed into the choice of models that scientific and normative aspects intertwine. It should be expected that, if Leopold got it right, an adequate concept of sustainability must be normative, dynamic, and multilayered. In subsequent chapters, I explore extant, disciplinary conceptions of sustainability and consider whether these existing approaches to understanding and defining sustainability can comprehensively capture Leopold's wisdom and whether they can guide us toward a more integrated—and normative—conception of sustainability.

As noted in the preface, this book is organized into two parts, each associated with a distinct task. Part 1 has mainly negative and preparatory tasks, as it distinguishes two possible paths toward sustainability and criticizes the one that implicitly, or explicitly, guides many theorists. The goal of this part is to show that a common way of thinking about sustainability—as a future state that has some characteristics deemed desirable, according to a chosen theory of value—ignores the depths of complexity involved in real environmental problems. Here, those theories will be criticized as placing unreasonable demands on any attempts at sustainable living, because actors lack the knowledge and foresight necessary to define sustainability goals in advance. In the face of uncertainty, it will be necessary to adapt our way toward sustainable policies and sustainable styles of living, and this recognition reveals a more positive contribution of part 1. By shifting from means-end thinking about how to reach some predefined point called "sustainability" to a more realistic project of improving decision making through experiment and adaptive action, it is possible to take a fresh look at the challenge of sustainability. Focusing more on decisions and less on theories of value opens up new areas for exploration and research on policy processes that work and also on those that don't work so well.

The search for sustainable environmental policies can be seen as a search for paths toward robust adaptation, understood by analogy with natural evolution and involving conscious learning. So, in addition to criticizing a common approach, part 1 introduces a new set of action-oriented concepts: procedures, adaptation, learning by doing, and others. These concepts and ideas will be applied only after an acute examination of the nature of environmental problems and decision contexts given in chapter 2. This examination will reveal that most environmental problems are wicked problems, which is to say they have no definitive solution and that they involve special and intimidating forms of uncertainty.

It is also to say that fresh assumptions and novel approaches will be needed, so chapter 3 introduces the methods and traditions of "adaptive management" as an alternative to traditional means-end rationality and the pursuit of a defined-in-advance conception of sustainability goals. Instead, adopting a realistic understanding of the problems faced, emphasis is shifted to improving procedures and improving the ways in which science can contribute to better policy directions; therefore, chapter 3 articulates the need for a new field of "evaluation studies"—a field based on the empirical study of processes of decision making available to communities seeking to act cooperatively and adaptively on the path to sustainability.

Part 1 culminates, in chapter 4, in a deeper analysis of the contest over the meaning and applications of the sustainability as a concept. Employing the concepts introduced in the earlier chapters of part 1, chapter 4 explores conflicts between economists and ecologists and introduces a definitional schema—a template for more specific, place-based definitions of sustainability appropriate to actual places and real community efforts.

Part 2, building on the advice of Adapt, addresses the more positive task of defining an approach to sustainability that assumes a more pluralistic language and a more flexible approach to valuing environmental change. Sustainability can then be thought of as finding a reasonable balance among plural values, while the goal of both sustainability studies and policies should be to make better decisions, ones appropriate for both short-term and long-term goals. Better decisions, chapter 5 will argue, can only emerge from a process that encourages deep deliberation in search of a "public interest." The study of such processes introduces a general method that will guide the process of seeking better and more sustainable decisions. Chapter 6 addresses the role of disciplines and specialized discourses in the search for better decisions.

Chapter 7 takes a more practical look at the tools available to enrich discourse and public deliberation and also provides opportunities for transformative thinking. While algorithmic solutions to environmental problems are impossible, given the wicked nature of most of these problems, there are better and worse processes by which communities can make decisions, and very useful tools can be employed to encourage creative, transformative thinking. The key to transformative thinking, developed in chapter 8 and based on Leopold's transformation, "learning to think like a mountain," is to entertain and explore multiple metaphors.

Chapter 9 examines the empirical literature about what has worked—and what has not worked so well—in actual community efforts at developing sustainable policies through deliberation and cooperation. Finally, chapter 10 explores the difficult problems involved in "scaling up" from local problems and local processes to larger scales, including global ones. While conceptualization and institutionalization of global change are not yet well developed, following Leopold's lead will provide an example of how to break free of unitary, dogmatic assumptions and how to evaluate patterns as they unfold in larger, normally slower-changing, systems. The next step, then, in scaling up toward global sustainability will be to learn to "think like a planet." Finally, the epilogue explores the implications of the book for the future of "policy analysis": overemphasis on specific methods of policy analysis can lock thinking into narrow channels, in service of narrow values. What is needed is transformative thinking—the proposal of new strategies that involve the integration of broadly pluralistic values, rather than preoccupation with methods more likely to yield precision than insight.

CHAPTER 2

The Decision Context

2.1. Two Orientations: Theoretical or Practical?

Some of the oldest conundrums in philosophy derive from the contrast drawn between theory and practice. Our personifications of Optim and Adapt exemplify radically different responses to the theory/practice contrast, as illustrated in figure 2.1. They take on, one might say, different orientations toward environmental problems.

Optim approaches environmental problems by first trying to develop and articulate a general theory of environmental value and, armed with such a theory, he attempts to articulate an ideal state of sustainability as a goal for the future. The theory, one can note in its favor, provides Optim with a general understanding of problems, guides him toward asking certain questions, and helps him design research projects appropriate to maximizing the values that are central to his theory. Optim and thinkers like him approach particular environmental problems by first generating a theory of environmental value and then applying their ideal theory as a yardstick to measure progress toward the ideal. He and his ilk are acting with a theory that implies there is a best solution to problems, and they pursue that theory in all situations to identify an ideal state. One can say, then, that they pursue an ideology—a set of general beliefs or assumptions that are effectively immune to refutation by empirical evidence.

Adapt, by contrast, is wary of grand theories derived before her careful examination of cases, and she tends to pursue understanding by precise ob-

What approach to evaluating environmental change will be most useful in actual situations in which communities face conflicts about what to do, especially when they attempt to pursue sustainable development?

What is the correct/best theory of environmental value, what sustainability objective does it imply, and what methods does this theory require or support?

Figure 2.1. Optim and Adapt evaluating change: Optim aggregates individual values and seeks to maximize one or a few variables; Adapt urges communities to experiment and deliberate in search of values they wish to project into the future.

servation and experimentation done case by case. In this chapter, in keeping with the resolve to follow Adapt and see where she leads, I will develop one important aspect of Adapt's strategy: paying close attention to the particular context in which problems are addressed.

The contrast between theory and practice, it is important to note, is by no means an all-or-nothing affair, which we signal by speaking of different orientations toward problems. Theoretically oriented scholars believe that their theories will eventually be made applicable to cases and that these theories provide guidance to the solution of particular problems. While more practice-oriented researchers and activists pay attention to particular problems, developing detailed knowledge of relationships exemplified in the particular case, they also pursue theoretical understanding that can be developed inductively, generalizing from detailed knowledge of multiple cases. So, when one refers to the theory/practice contrast, it is best not to think of these poles as exclusive of each other, but rather as referring to differing orientations toward similar problems and similar bodies of knowledge. Optim starts with theory, with the intention of eventually applying it to particular cases. But, following Adapt's strategy of paying attention to cases, we focus more on day-to-day decisions and the contexts in which they are made and less on theory-driven speculation about possible ideal future states. In particular, this chapter will examine the decision contexts in which most environmental problems are addressed.

2.2. Decisions in Environmental Conflict Situations

Often in discussions of environmental decision making, situations are conjured in which a policy or action has been proposed and discussed and is coming up for a decision point, such as a legislative vote or an executive decision to be made and announced on a given day. Surely these situations do occur, and when they do, they can be paradigmatic of environmental decision making; most writers on decision theory and practice discuss decision making in ways appropriate to these contexts in which there is a vote up or down, or an anticipated policy announcement from an administrator. In such contexts it may not seem out of place to think of decision makers as toting up costs and benefits or as identifying objects of intrinsic value whose values trump other instrumental values. In fact, however, these situations show only the tip of the iceberg of decision situations. The decision process also includes all the smaller-gauged decisions that take place when committee members gain support to insert new language into a bill, when committee members allow lobbyists to insert loopholes in laws, and when administrators decide to—or not to—hold hearings, and so on, as well as those decisions that administrators make covertly (O'Neill et al., 2008).

Thus it is tempting but misleading to focus only on final up-or-down votes or to think of environmental problems as having the form of a well-formulated question on which one can focus to gather evidence about the consequences of specific choices. Such a narrow focus would cause one to miss informal forces that implicitly affect decisions and decision making. Decision making is a process: therefore focusing solely on the public and obvious decision points can miss what is most important in shaping policies—the informal and ongoing process by which policy is developed. The antidote to this oversimplification is to think of policy development not as represented in isolated decision events but rather as an ongoing, iterative process that perhaps is punctuated by important decision points.

Scale is another issue in public discussions about environmental problems. For example, confusions of scale pervade the discourse, a topic that will require careful discussion and analysis throughout this book. For example, a discussant might say with great feeling, "The government should do something about X." But it makes a great deal of difference if the citizen invokes local government or the federal government than if the social, economic, and political forces thereby engaged take steps to determine the scale of the proposed action. Problem statements themselves suffer ambiguity, as when individuals agree that a body of water is "polluted," though they may bound the water body differently (e.g., decide to include or exclude tributaries), which leads to participants talking past each other, creating a stalemate. In other cases, problems in their physical manifestations do not match the territorial jurisdictions of governments, as when bodies of water such as the Great Lakes are being polluted from both sides of state or national boundaries.

I first came to see the centrality of scale one time when I talked with a water manager who oversaw water provisioning from Lake Lanier, the large impoundment north of Atlanta, which is a major source of drinking water in that region and which also generates electricity and controls floods. Doing research on how problems get formulated, I was trying to understand the vitriolic and often litigious relationship between counties surrounding the lake, the Lake Lanier Association, several municipalities, and the Army Corps of Engineers, all of which comanage the lake. One thing was clear from reading the newspapers: among residents, one strong emphasis was on occasional outbreaks of E. coli and other bacteria. Residents connected to sewage systems felt certain that those with septic tanks are ruining the lake, while those with septic systems claimed the lake was being polluted by ineffective sewage treatment plants. We were standing at a boat landing at lakeside when I asked the water expert what he thought of the residents' concerns. His look turned disdainful and, sweeping an arc with his arm large enough to encircle at least a dozen Canadian geese in sight, he said, "If the residents want to reduce E. coli contamination, I suggest they get together and have a goose roast."

The point of this example comes home if you compare this controversy about localized E. coli outbreaks with the concerns of a colleague of mine who is a limnologist (one who studies bodies of freshwater), from whom one gets another story. He remarked, "Yes, Lake Lanier is being polluted and sewage treatment might play a small role, but the real problem is the hundreds of chicken farms dotting the landscape north of the Lake." It seems that chicken farmers spread manure too heavily on their land, and after a heavy rain, all that phosphorus and nitrogen pours into Lake Lanier. "The sediments are loaded with phosphorus," he continued, "so if any event disturbs that phosphorus and discharges it into the water column, the lake could turn into an anaerobic slime pond." So, here we have two different perspectives on the problem of polluted Lake Lanier: what is interesting is that each view of the problem can be related to quite different spatial models and contrasting ways of bounding the problem. Setting boundaries around the scope of a problem is almost always a key aspect of problem formulation.

Spatial ambiguity is only one example of the ways in which public discourse about environmental policy is often beset with confusion and poor communication. Such confusions occur in many intractable public conflicts, where the disputants do not actually view the same problem. In these situations, cross-talking among interest groups often results in more litigation than communication, cooperation, and resolution.

2.3. Most Environmental Problems Are Wicked

If a focus on theory risks too much, then empirical attention to cases of environmental conflict informs us that, in most cases, it is difficult to find clear-cut choices and decision points. As disputants define and redefine the problem, participants are hard pressed even to agree about what they disagree about.

One dramatic statement of this point was provided by a leader in the field of operations research, Russell Ackoff (1979), who says, "Managers are not confronted with problems that are independent of each other, but with dynamic situations that consist of complex systems of changing problems that interact with each other. I call such situations messes. Problems are extracted from messes by analysis. Managers do not solve problems, they manage messes." The lesson: it is necessary to problematize problem formulation itself.

A similar point is also well made by the systems analyst and futurist Donella Meadows, who notes that, in contentious situations, discussants usually define a problem as the lack of their favorite solution, rather than providing a thoughtful analysis of the values threatened and the reasons those values are underproduced or underprotected by the currently functioning system. People say, "The problem is, we need to find more oil," or "The problem is, we need to ban abortion." She suggests, "Listen to any discussion, in your family or a committee meeting at work or among the pundits in the media, and watch people leap to solutions in 'predict, control, or impose your will' mode, without having paid any attention to what the system is doing and why it's doing it" (Meadows, 2008, p. 171). In such situations, no clear view of the "real problem" will ever emerge, nor can formal, rational analysis, however sophisticated, be brought to bear on "messy" situations. Note that the "predict, control, or impose mode" of which she speaks is exactly the strategy one would expect of Optim, once he has defined the objectives toward which sustainability policy is directed.

The point Ackoff and Meadows make was also stressed and developed, using different terminology, in one of the true classics of decision theory, a profound little paper called "Dilemmas in the General Theory of Planning" (Rittel and Webber, 1973). Its authors, Rittel and Webber, distinguished benign from wicked problems. Benign problems include, for example, mathematical problems as well as certain scientific problems, which are benign in the sense that they have a single determinate, correct solution. If one finds the answer to a benign problem, the solution will be accepted, on demonstration, by all. Wicked problems, by contrast, involve situations in which the problem formulation itself is controversial; these situations represent, as Ackoff said, "messes." In their discussion of wicked problems, Rittel and Webber mention ten characteristics of wicked problems, which appear in box 2.1.

Box 2.1. Rittel and Weber's Ten Characteristics of Wicked Problems

1. There is no definitive formulation of a wicked problem.
2. Wicked problems have no stopping rule.
3. Solutions to wicked problems are not true or false but are either good or bad.
4. There is no immediate and no ultimate test of a solution to a wicked problem.
5. Every solution to a wicked problem is a "one-shot operation"; because there is no opportunity to learn by trial and error, every attempt counts significantly.
6. Wicked problems have no enumerable (or an exhaustively describable) set of potential solutions, nor is there a well-described set of permissible operations that may be incorporated into the plan.
7. Every wicked problem is essentially unique.
8. Every wicked problem can be considered to be a symptom of another problem.
9. The existence of a discrepancy representing a wicked problem can be explained in numerous ways; the choice of explanation determines the nature of the problem's resolution.
10. The planner has no right to be wrong.

Rittel and Webber suggest, without explicitly saying so, that wicked problems occur because, when a group or a community faces a problematic situation, individual members bring different values to the discussion and, given the diversity of values and perspectives, it is difficult for all to agree on how to remediate the problem. Indeed, coming from different perspectives, and pursuing different and sometimes competing values, groups facing a wicked problem cannot agree how to formulate the problem at hand. By extension, when environmental managers address problems that confront them, what they actually face are usually "messes," and the problem situations confronting them almost always involve wicked aspects. Environmental problems are, at their core, wicked problems. They represent a range of discomforts expressed in the public discourse, but more careful analysis shows that those discussing the problems actually formulate the question in ways that create confusion and miscommunication.

The Rittel and Webber distinction between benign and wicked problems has proven so popular because it provides an explanatory hypothesis for a frustrating phenomenon that affects many people—policy gridlock over huge, important social problems like water pollution and reduction of fossil fuel use. Once most people are made aware of this distinction, they nod their heads and think immediately of a frustrating example in which public discourse on a local or regional problem generates much heat but little light, and they recall certain interest groups working at cross purposes to block all constructive action to address the problem. "Ah!" they say, "it's a wicked problem" (or they would say that if they knew the term). Merely listing the characteristics of wicked problems and articulating a category that captures them causes a glint of recognition, a sense that one now sees the source of one's frustrations.

Such a classification, of course, gets us no nearer any solutions; indeed, by encouraging a fatalistic attitude, it might create policy lethargy and acceptance of an unacceptable status quo. Is it possible, though, to fight fatalism and to struggle to understand how to move forward toward solutions, before a clear formulation of the problem has emerged?

The distinction between benign and wicked problems is vital in this book, but not because it brings comfort in the face of failures. Better understanding the characteristics of intransigent problems can offer guidance in developing more effective processes, processes that can lead to an emergent concept of sustainability. Far from inferring from the judgment that a given problem is a wicked problem and therefore is hopeless, we adopt the practical goal of articulating several strategies—based on what John Dewey called "social learning"—and we suggest that many conservation challenges associated with moving toward sustainability will require profound changes in how we think and act (Dewey, 1927). This goal is to provide guidance—not toward a specific goal or objectives associated with a preconceived optimal state, but instead guidance in asking the right questions, avoiding conceptual confusions, and constructively addressing wicked problems by developing best practices for public processes of deliberation, debate, and social learning.

—✳—

In chaotic situations, it often helps to gather a series of understandings and recommendations that can guide discussion and action—what can be referred to as "heuristics." Heuristics can bring some structure to chaotic situations by focusing attention on specific aspects of a problem's wickedness, encouraging reasonable responses even when one is facing a truly wicked problem.

I do not counsel the fatalistic acceptance of gridlock when facing wicked problems. Instead, I recommend that the determinants of wickedness be determined and ways found to ameliorate them. In the following text, I will introduce heuristics that can show a way toward progress in addressing wicked problems; these heuristics are summarized in the appendix. Rittel and Webber never provided a general definition of wickedness. They cagily relied on a list of shared characteristics and examples. As one looks at the characteristics more analytically, however, it is clear that the characteristics emphasize different aspects of wickedness. Once these aspects are separated, each aspect may be open to clarification and perhaps to partial amelioration. By showing that "wickedness" does not refer to a monolithic, impenetrable mess, the stage is set for some advice from Adapt, who may be able to suggest ways to deal with each aspect of problem wickedness that afflicts them.

In my earlier treatment of wicked problems elsewhere (Norton, 2005), I noted that Rittel and Webber's list of ten characteristics is somewhat redundant and also that the characteristics tend to cluster around several distinct themes. A closer look at the list of characteristics allows a grouping of aspects of wickedness associated with these themes. Separating the themes may guide the development of heuristics that will help address some of these aspects, improving the odds of progress on wicked problems. The characteristics of wicked problems on Rittel and Webber's list can be separated into four subgroups (Rittel and Webber, 1973; Norton, 2005):

1. Problems of problem formulation (characteristics 1, 2, 3, and 9)
2. No computable solutions (characteristics 2, 4, 6, and 9)
3. Nonrepeatability (characteristics 5, 7, and 10)
4. Temporal open-endedness (2, 4, and 8)

All ten characteristics are symptoms of underlying value conflicts, which ultimately contribute to the complexity of the cases. This does not mean, however, that one cannot learn something by examining the exact ways in which conflicts are manifest in wicked problems. Perhaps paying attention to specific characteristics of wicked problems as they present themselves in messes can contribute to agreement and cooperation, even in the messiest of situations. This is consistent with Rittel and Webber's observation that wicked problems, while never susceptible to a final solution, can sometimes be "resolved" by temporary solutions to certain aspects of, say, a complex environmental problem, producing improvement without resulting in a comprehensive solution. With this in mind, now let's explore the four subgroups in more detail:

1. Central to Rittel and Webber's understanding was the problem of problem formulation itself. It is in this central characteristic that the problem of value-laden problem formulations creates contexts in which individuals and groups, seeking cross-cutting goals and diverse values, fail to see the same problem, even when they are exposed to the same body of descriptive data. I will say no more on this theme here—it is fully explored in Norton (2005) and as one of the ongoing themes of this book.

2. Several of the characteristics of wickedness have to do with the noncomputability of solutions to wicked problems. This aspect of wickedness is perhaps best expressed by Rittel and Webber when they say that formal decision models become operational "only after the most important decisions have already been made" (1973, p. 162). I have taken this theme to force a major shift in the way we understand both policy analysis and the processes of policy formation and implementation; this shift will be detailed in chapter 3.

3. The nonrepeatability theme, expressed, for example, in characteristic 5 ("Every solution to a wicked problem is a 'one-shot operation'") emphasizes that there will be no one-size-fits-all solutions and that the complexity of situations entails that each possible resolution is unique and applicable in only one place. This theme is important, since it directs attention to communities facing unique situations that are determined by both social and ecological contexts.

4. Temporal open-endedness, while frustrating to those who expect closure, can be both emblematic and illustrative of useful, multiscalar models (Norton, 2005). Rittel and Webber provide the following comment on difficult decisions exemplifying characteristic number 4: "The full consequences cannot be appraised until the waves of repercussions have completely run out, and we have no way of tracing the waves through all the affected lives ahead of time or within a limited time span" (1973, p. 63). They capture, in this theme, the essence of the sustainability problem. One can analyze any phenomenon from many viewpoints and from multiple perspectives; there is no temporal limit after which our activities will have zero impact.

Thus the problem of sustainability is inherently open-ended. Systems that might capture and measure sustainability must be open systems, and this openness to outside forces creates yet another cascade of uncertainty affecting every smaller scale.

2.4. Strategies for Living with Uncertainty

At the heart of the difficulties in dealing with environmental problems is the magnitude of the uncertainties typically associated with them and the four themes extracted from the list of characteristics of wicked problems. These characteristics are indicative of some of the huge uncertainties involved in addressing messes. The unavoidable conclusion of the last section, however, was that messes have to be addressed in contexts of great uncertainty.

Many experts have discussed uncertainty and how to deal with it, but despite these discussions, uncertainty dominates many decision situations. Nobody has a foolproof way of dealing with our pervasive ignorance when facing complex problems. Commentators have tried to manage the "problem" of uncertainty by identifying several types of uncertainty. Ignorance, when recognized, was for Socrates the very font of wisdom, in that acknowledged ignorance can be the catalyst for learning. It is also important to realize that some uncertainty is simply epistemological—there is not enough information about a phenomenon being faced—while other uncertainty is a result of complexity so deep that, even in principle, the outcome of a process is unpredictable.

Different kinds and versions of uncertainty and ignorance have been taxonomized, but the goal of conquering, or even "cornering," uncertainty by packaging it in comforting categories is not possible in the face of wicked problems. These problems arise and persist in practice when affected parties bring multiple conflicting viewpoints and values to a discussion of what should be done. Uncertainties of many types arise when there is no agreement on problem solutions because there is no agreement on the proper formulation of the problem at hand. The word "uncertainty" is typically used as a catchall term for what one wishes one knew but does not. The problem of uncertainty, then, cannot be solved, nor does it help much to put it into boxes with varied labels. What is necessary is a method that accepts uncertainty as an unavoidable element in all complex decision situations and that develops a strategy for living with it.

Optim, remember, approaches sustainability problems by developing, based on a broad theory of value, an objective that would represent achievement of sustainability. Given this strategy, Optim's approach relies heavily on seeing a sharp dichotomy between the objectives or ends sought and the means necessary to achieve the objectives.

The means-end distinction has been prominent in conservation theory from virtually the beginning, as when Gifford Pinchot, the first head of the US Forest Service, understood forestry to be a profession guided by social objectives such as maximizing output of harvestable timber, objectives that he took to be justified by public opinion and political decisions, essentially providing foresters with values and goals determined from outside their profession and practice. This approach dominated environmental management and conservation from its beginnings in the American Progressive movement late in the nineteenth century. Based on the utilitarian thinking of Gifford Pinchot and his allies in forestry and other fields, an approach to management was developed that is often called scientific resource management (SRM). Pinchot characterized his philosophy of conservation as "the use of natural resources for the greatest good of the greatest number for the longest time" (Pinchot, 1947, p. 326). Consequently, for him, the professional forester was an expert in devising and testing means to ends that were determined outside forestry in a political process. This distinction allowed Pinchot to treat forestry as a scientific, hypothesis-testing profession, one that, like the military, took its marching orders from "civilians." The claim of the profession of forestry to expertise, and its sometimes-arrogant rejection of advice from nonprofessionals, depended on a claim to having special scientific expertise in describing how to grow trees efficiently, among other things. The professional manager's claimed expertise in means depended on a sharp separation of questions of fact from values; scientific forestry, then, was a means-oriented field, and trained, professional foresters should manage the forests to maximize achievement of goals set politically.

SRM (with a few exceptions, including Leopold) dominated thinking about environmental management until, in the 1960s and, especially in the 1970s and 1980s, strong, ecologically based criticisms called the whole approach into question. Since the 1970s, many ecologists and other scientists, while uniting around ideas such as ecological systems thinking, resilience, and the importance of biodiversity, realized that SRM could not be sustained as an ongoing approach to management. Ecologists and conservation biologists began to question the reductionistic and atomistic science that dominated resource management agencies, and they began experimenting with more systems-oriented thinking. In the United States, this process of criticism gained political notice because it corresponded to a major demographic shift as people moved from city centers to suburbs and as families with growing incomes began to demand outdoor recreation and protection of aesthetic aspects of the environment. These changes, in turn, put pressure on agencies such as the Forest Service to pay attention to broader social values than simple maximization of timber production.

While much of the innovative thinking that occurred during the second half of the twentieth century was ad hoc and not theoretically unified, by the late 1970s the insights of the critics of SRM began to coalesce into a movement that came to be called "adaptive management." Adaptive management has been, and remains, a reaction against and a replacement for the tradition of SRM. While both approaches emphasize science, their notion of science is quite different: For Pinchot, timber production to support human development and welfare was the unquestioned goal of forestry. The role of science, he thought, was to determine the correct means to achieve that social goal. Adaptive managers react both against the narrowness of the social goals assumed and also against the sharp separation of means from ends.

SRM professionals embraced two key tenets that ensured that they, in the tradition of Pinchot, would conceive and address resource problems while ensuring a sharp separation of considerations of means and ends:

1. The management technique claimed for science a form of objectivity by which systematic use of the scientific method could, in principle, eliminate error and, based on empirical evidence, provide a correct description of the physical world.
2. By pushing the question of goal-setting and evaluation of changes out of fields like forestry science, agronomy, and mineralogy, questions of value were isolated and considered unscientific and "subjective."

The first tenet allowed resource managers, schooled in the use of the scientific method, to claim expertise once clear goals, such as maximizing the board feet of marketable timber, were set for them. After the management goals were set, expert resource managers would be available with the expertise necessary to "deliver the goods." The second tenet, on the other hand, reinforced a separation of value considerations from factual content, and a belief that values are "subjective" and unscientific pushed discussion of goals and values into the background. Scientific resource managers thus deferred to economists and politicians regarding ends and goals, which allowed them to concentrate on the scientific development of means to produce goods supporting human welfare. This also supported their claim to be experts— experts in growing trees or in efficiently using water. Scientific resource managers, one might say, were embodiments of the philosophy of Optim.

In chapter 6 I will return to the many issues raised by tenet 2. Here, our target is tenet 1 and the role it assigns uncertainty within processes of management. For Pinchot and the scientific resource managers who followed in his footsteps, the sharp separation of means from ends allowed them to claim scientific expertise in resource management decisions and to insulate their decisions from political and public scrutiny in many situations. Scientific resource managers' understanding of uncertainty as regards only means to ends supports the general strategy of treating the search for sustainability as the pursuit of objectives set outside science by economists and politicians. It also allowed them to act as if the problems they faced were benign problems, likely to yield to applications of the scientific method.

In other situations, where action is guided by past practice, when challenged about their lack of knowledge, scientific resource managers reply that, while they do not have all the knowledge of means that would be ideal, such knowledge can be attained—that, *in principle*, the solution to uncertainty is more scientific funding and more scientific research. The key (and misleading) phrase here, however, is "in principle." When faced with mountains of uncertainty, scientists maintained an optimistic outlook and professional demeanor. While their ignorance is huge, their past success and their faith in the future of science justified them in operating as if obtaining more success in sorting hypotheses into true and false groupings was just a matter of time, and they believed that the march of science would eventually fill in the gaps and, in principle, every empirical problem would yield a unique solution. This bit of collective self-deception has been so pervasive and so determinative of our thinking that it deserves a label: I'll call it the "in principle patch" (IPP). Whenever there is a hole in our knowledge— something we would like to know so we can act intelligently—Optim and the other advocates of SRM invoke the IPP: "We may be ignorant," they say, "but it is only a matter of time before science will eventually provide the necessary knowledge. Certainty is just around the corner. . . . P.S. Send more money for research!"

The crucial point here is that the IPP is applied by scientific resource managers to paper over gaps in what they consider to be "objective scientific knowledge" of means, all the while ignoring the real gap in their strategy: the scientific method, as they apply it to questions of means only, cannot resolve the value differences that make complex social problems wicked. To the extent that what to do in wicked-problem situations is uncertain because of their wickedness, the IPP as applied to scientific knowledge of means will always be too small to fill the gap between what one knows and what one wishes were known in order to act intelligently.

Uncertainty, both to Optim and to scientific resource managers, seems like an obstacle standing in the way of a computation of a best outcome according to an objective that was set in advance. Recognition that most environmental problems are wicked problems undercuts faith in computation and broad generalizations and shifts attention to more holistic and systematic thinking. As SRM came under more and more devastating criticism, ecologists, ecological economists, and others began pursuing an alternative, more pragmatic approach to management. In the United States, this movement has come to be called "adaptive management," which to some extent has penetrated all levels of environmental management. So christened in 1977, adaptive management represents the most promising attempt yet to gather and systematize, to the extent possible, the fragmented research and thinking that has constituted the ongoing revolution that is replacing SRM with a more systems-oriented, pluralistic, and democratic approach. I begin to develop the core ideas of adaptive management in the next chapter.

CHAPTER 3

A Brief Philosophy of Adaptive Ecosystem Management

Adaptive management has emerged as the main competitor of and successor to scientific resource management (SRM). Many commentators on environmental policy fail to notice that environmental problems are wicked problems, so they continue to advocate ineffective approaches and methods to complex problems. What is needed is a set of methods that will work in messy situations in which confusion about objectives is an essential aspect of the challenge faced.

No algorithmic and computational systems of analysis exist that can comprehensively model or even adequately represent complex sociopolitical problems (Zia et al., 2012). Here, I begin to fulfill the promise of the preface to clear away the conceptual underbrush by dismissing certain intellectual quests as hopeless—in this case, the dogged pursuit of a computational solution to complex problems.

What are the currently available approaches to evaluating ecological and environmental change? To pursue sustainability, one will need some kind of method or technique for evaluating actual and proposed changes in the world, especially changes to which humans can contribute by virtue of their policies and actions.

Some theorists find the argument of Rittel and Webber an unacceptable outcome when they realize that the "in principle patch" (IPP) is never big enough, or the right shape, to cover our knowledge gap. Not only will none of the formal models proposed allow a comprehensive analysis, it further seems to them that recognizing the pervasiveness of wicked problems is actually a disaster; if one encounters wicked problems on the way to achieving sustainability, one will never find a rationally justifiable way toward it. Before exploring that question, I will underline the negative conclusion that there are no computational solutions to complex social problems, thereby clearing the way for a more realistic assessment of the methods and tools available.

3.1. The Case against Computation

Closely allied with Optim and his ideological approach is a crucial myth: that a computational solution exists for every complex problem. Captives of this myth harbor, against all evidence, the hope of creating a theoretically guided model that will allow the computation of a set of unique objectives for sustainability, capable of focusing the search for means to those objectives. For example, economists, when touting economic growth and efficient fulfillment of human preferences, provide computational models that support unlimited consumption.

Before proceeding in more positive directions, I resoundingly reject this myth, explaining how that myth confuses the issue and directs attention away from hopeful alternatives. The myth is highly influential, and controversial, in decision analysis. Methods that embrace decision analysis as a search for algorithmically generated (computational) results identifying a single best solution include the following:

- Rational choice models
- Microeconomics; cost benefit analysis as a decision tool
- Optimizing programs in operations research (e.g., linear programming)
- Competitive game theory

Whatever their usefulness as partial guides and useful analogies, these approaches and their equations cannot offer a complete systemic model, or a complete analysis, of any complex social problem. Attempts to analyze wicked problems in their messy state will always spill over the conceptual boundaries of any purportedly complete system for analyzing problems. We can now begin to understand the profundity of the problems with Optim's approach. Ackoff's version of wickedness, which contrasts problems (labeled benign) and messes (labeled wicked), says that managers face messes, not problems, and that problems are extracted by analysis from messes. (I see only a semantic difference here, in comparison to Rittel and Webber's distinction.) Accept Ackoff's terminology, but then ask: by what method of analysis can one extract problems from messes, which Ackoff suggests is the first step toward progress? This poses a quandary. Reaching a fruitful formulation of a problem involves simplifications—since all disciplinary models are, at best, partial pictures of reality—but employing any disciplinary model of analysis will involve assumptions associated with that model and its home discipline. So, using Ackoff's phraseology and reasoning, we need a form of analysis that will get us from messes to problems, yet the consensus expressed by Rittel and Webber and by Ackoff apparently implies that any such method will oversimplify and, in the process, fail to capture the full dimensions of the original problem.

Therefore, with respect to wicked problems, we face an analytic void. In the process of taming a wicked problem, messes are simplified in ways that determine which discourse and which analytic tools will be used. The methodologies themselves, however, cannot justify their own use based on resources internal to the model. Hence, when one addresses wicked problems, any choice of methodology must appeal to considerations external to the chosen model. In this world of messes, special interests, politics, and power often skew the formation of problems before they reach the policy agenda. For example, applying an economic model to problems of land use often causes decision makers to ignore the resilience of ecological systems, a characteristic that is not easily incorporated into a model based on individual preferences.

Approaching the problem from the viewpoint of policy analysis and applied decision theory, it is possible to formulate the general result outlined above, as follows: for any complex public controversy of importance to diverse interests, it will never be possible to model that conflict in such a way as to provide a computational solution to it (Zia et al., 2012). So what should one do in response to this hard-won conclusion? What should be the reaction to the end of the IPP era in thinking about decision making? The Holy Grail of automated decisions is wholly myth. The IPP could not, even in principle, fill the void. Open systems always interact with an outside world, meaning that any complete analysis must extend beyond the system boundaries; this explains the spillover metaphor introduced above. In an open system, any problem will have effects—and causes—beyond the system specified. Complex, dynamic systems are necessarily open. They can maintain themselves only by exchanges with larger systems, including absorption from, and dissipation of, energy into the environment. Thus, environmental problems will always exist within complex, many-layered systems, and those systems will look very different, depending on the perspective from which one views them. If one puts humans into such a system, placing them differently so that they will have different perspectives on the workings of the system, the resulting society will face wicked problems.

Whether a systems analyst endorses computational solutions will be closely tied to the choice of whether to envision the system models created as open or closed systems. Closed systems are assumed to contain all the factors affecting behavior of that system. At the heart of all these arguments against computational approaches to analysis is a recognition that attempts to sharply separate empirical discourse of the sciences from evaluative discourse must ultimately fail. The analytic frameworks one chooses, however sanitized of value claims and implications, rely on choices made outside the closed systems posited in order to formalize the decision process. Those choices inevitably embody assumptions that are, in turn, supported by underlying, often unnoticed, values. As recognized by Rittel and Webber and by Ackoff, it is often the case that invisible decisions about the methods to be used in the analysis cannot themselves be conclusively supported within any system of analysis. Messes can only be turned into problems by a kind of analysis that cannot, itself, be formalized.

If we are to have a useful criterion to determine success and failure of experiments, we must have an accepted method of evaluating changes to environmental and ecological systems. But how should we evaluate the impacts of a decision that improves human welfare on the scale of individuals but that portends changes to large environmental systems that may be negative and irreversible?

Fortunately, recognizing that comprehensive computational approaches are impossible has not left all analysts stymied and unable to engage in rigorous analysis of situations and applications to policy. Some advocates of general systems theory (GST)—despite the fact that this field of study originated in unjustified optimism regarding opportunities to compute best outcomes—came to recognize the impossibility of comprehensive formal analyses. This shift, concurrently with the decline of positivistic and objectivist concepts of science and analysis, encouraged important innovations. Critical systems theory (CST) has developed rigorous methods that, even when working within the confines of a defined system, can recognize the limitations of assumptions of dominant models of complex systems. (See box 3.1.)

Especially relevant to our discussion here is the work of Werner Ulrich, who engages in a type of analysis called "boundary critique." When supplemented by boundary critique, CST can, on the basis of information that an important viewpoint is excluded from a model, question whether the boundaries of a system of analysis are drawn too narrowly. Noting that, in many cases, the dominant representation of a system can exclude factors important to those out of power, Ulrich argues that a broader systems analysis—one that includes a critique of the boundaries set for the analysis—can "emancipate" the analyst from too-narrow formulations of a problem and from inadequate models for addressing the problem. Ulrich's work, deserving of much more attention than it receives, can only be introduced here, yet his line of reasoning provides considerable promise for developing noncomputational but nevertheless rigorous methods of analysis (Ulrich, 2003).

Box 3.1. Werner Ulrich and Critical Systems Theory

CST is one strain of systems theory, the strain that concentrates on decision analysis and how systems learn through feedback to develop managerial and decision-making models. Werner Ulrich criticizes traditional CST as reluctant to recognize that a true critique cannot be simply methodological in nature. His "boundary critique" approach incorporates CST but also allows the system to learn, for example, that an important viewpoint is excluded because the boundaries of the system have been drawn so as to exclude affected parties from the analysis.

Ulrich concentrates on the scientific and intellectual context of the decision process, using scientific information to force questions and open up new lines of inquiry. He thus focuses on the epistemological aspects of the management problem; the contextual constraints he explicitly addresses—including the limited role of algorithmic solutions—are limits to the epistemological process of iteratively improving policy through learning.

The strength of Ulrich's approach is that, while he renounces the algorithmic dream, he nonetheless sets out to develop a rigorous and comprehensive decision procedure, which leads to critiques of the results implied by the setting of inappropriate or even illegitimate boundaries. Ulrich's work thus recognizes that setting problem boundaries too narrowly can lead to unfair exclusions of interests. For him, any assumption that excludes viewpoints or significant information from the decision process is seen as a "boundary problem."

3.2. The Current Situation in the Field of Evaluation Studies

The title of this section is a joke or, what in philosophy of language is called a "nonreferring expression." There *is* no current situation in evaluation studies, because there is no unified field that could reasonably be called that. As in a failed political state ruled by a number of independent and antagonistic warlords, the task of evaluating changes in environments, whether before or after they occur, is undertaken as guerrilla warfare. Because members of each separate group have been trained in the same disciplines, with each using their own concepts and measures and each demonstrating intellectual, if not personal, antagonism toward members of other groups, conflict is inevitable. Such antagonism might be surprising, since the task of evaluating change can be stated as a simple-sounding question: when change occurs in an environmental system, how can it be evaluated (either prospectively or retrospectively)?

As it turns out, answering this question has been anything but simple, at least partly because the answers actually proposed have been influenced by differing ideologies and associated theoretical frameworks that give rise to conflicting ways of describing and evaluating change. Interdisciplinary discussions of environmental value thus unfold as turf wars among different disciplines. While one might list several groups that exemplify this fragmented methodological approach, it is possible to illustrate the barriers to creating a unified approach to evaluating change by focusing first on two theoretical disciplinary approaches to understanding natural value: "Economism" and "intrinsic value theory" (IVT).[1] We shall see that both these theories encourage an emphasis on narrow values, much as Optim does, and each exemplifies the confusion and polarization that drives the fragmentation of methods proposed and used to evaluate environmental change.[2]

This fragmentation exists because environmental economists and environmental ethicists have adopted different, and apparently incompatible, theories of environmental value, as illustrated in table 3.1. These opposed theories of value block the development of a unified field of environmental evaluation and lead to polarization in the public discourse about environmental values.

The field of environmental ethics has been largely dominated by a coalition of otherwise diverse thinkers who describe themselves as "biocentrists," as "ecocentrists," or simply as "nonanthropocentrists," and who often describe themselves collectively as "radical environmental ethicists." These philosophers, who disagree on much else, are unified in sharing the belief that it is essential to distinguish between two kinds of value: "instrumental" and "intrinsic." Intrinsic values are attributed to beings having a good of their own; historically, human individuals have been understood as the paradigm case of intrinsically valuable beings. Instrumental values, on the other hand, are values that are derived from humans' use and enjoyment of nature. Historically, in discussions of ethics, only human individuals are thus attributed "intrinsic value" (as having "a good of their own").

When nonanthropocentrists agree that not only humans but also other elements of nature have intrinsic value, they challenge any notion of human

Table 3.1. Comparison of types of environmental values

	Economism	Intrinsic value theory	Pluralism adaptive management
Type of value emphasized	Individual preferences (WTP) are aggregated to measure values	Intrinsic value of nonhuman elements of nature override instrumental values (preferences are trumped)	Pluralism of values. Multiple values and multiple ways of expressing values
Can values change?	No. Preferences are fixed and exist prior to actual choices	Yes. Values can change through criticism of anthropocentrism and adoptions of nonanthropocentric values	Yes. Adaptive management encourages deliberation and reformation of values through experimentation
How do values function in decision making?	Good decisions maximize preference fulfillment across whole society	Good decisions favor intrinsic value of nature over nonvital instrumental values	Through public deliberation, seek to find appropriate balance over full range of important values
What should guide decisions?	Cost-benefit analysis (CBA; plus judgment regarding nonquantitative values)	Protection of intrinsic value, wherever it occurs, to the extent possible	The emergence of a multigenerational public interest

exceptionalism, rejecting claims of special moral status for human beings. Advocates of this view enter the public debate arguing that the intrinsic values of natural objects have a special moral status that—at least in many cases—trumps the more everyday instrumental values that humans derive from natural systems. These environmental philosophers, far from joining economists in a common discipline, have made their priority the criticism and absolute refutation of economists' theories, especially Economism itself.

Further, this intellectual animosity can be seen as a direct result of conflicts that go to the heart of moral theory, pitting the dominant forces of economics and environmental ethics in a polarized conflict. Environmental economics, which emerged as a specialization in mainstream economics beginning in the 1960s and 1970s, adopted the value theory of neoclassical economics that was based on a specific form of ethical utilitarianism, according to which the maximization of human welfare is the goal of economic activity. Such activity is measured initially in terms of individual preference and then is aggregated with consumer surplus (what consumers would pay over what they in fact pay) to represent social welfare. Note that this Economistic viewpoint has three tenets:

1. *Anthropocentrism.* Only the good of human individuals counts in economic calculations.
2. *Utilitarianism.* Human value ("happiness") is constituted by the ability (income and assets) of individuals to achieve individual welfare.
3. *Aggregationism.* Good policy can be understood by aggregating individual values and aiming to maximize the overall welfare of the society.

Many philosophers, including some environmental ethicists, are utilitarians of various types, and many of them agree that the overall good of a society should be estimated by aggregating individual welfare. So the conflict between environmental ethicists and Economists that prevents the emergence of a subdiscipline of "environmental evaluation" is mainly over anthropocentrism. But this difference goes very deep, theoretically and ideologically. It goes far beyond differences in the estimates of particular evaluations. Advocates of the two approaches disagree about what has value and whose welfare counts. It's difficult to see a common discipline emerging without some consensus on which entities have value at all. There can apparently be no reconciliation of these two views as long as both assume a monistic stance—insisting that their view is the only correct one and that conflicting views are mistaken. I confront this conflict by rejecting monism in favor of a more pluralistic view.

Even more importantly, these differences in the practice of identifying and measuring environmental values affect public discourse. The development of competing theories and cross-disciplinary antagonism leads to multiple and competing methodologies, with advocates of each approach dismissing arguments and measures provided by the other discipline, causing confusion among activists and practitioners. These conflicting claims and heated critiques of alternative methods result in doubt and confusion even among well-intentioned participants in public processes. For example, economists typically assign little or no value to a species that has no human use, whereas nonanthropocentrists might attribute intrinsic value to the species.

From the broader viewpoint of the argument of this book, it should be apparent that these debates cannot be resolved on their own terms. Attempts to identify the correct theory of environmental value and its appropriate method for evaluating environmental change are likely to fail. It may be more productive to ask: What approach to evaluation of change will be most useful in actual situations in which communities face conflicts about what to do, especially when they attempt to pursue sustainable development?

Given the speculative nature of the general theories of value proposed by the contestants, these debates are formulated in a way that will not yield to empirical test. If someone says, "All environmental value is economic value— all environmental goods are commodities that could, in principle, be bought and sold in markets," there is no observation that will demonstrate—or refute—this claim. Its truth is built into the definitions of economic theory. Likewise, statements that a given species or system has "intrinsic value" are not associated with any observable and agreed-on observable mark. These ideologies are based on a simple definition that was developed when the conceptual frameworks of the Western tradition were far more dualistic and far more likely to draw sharp lines between inert matter, which can be used as one pleases, and objects with a soul. Trying to resolve questions and clarify distinctions that emerge from a now-obsolete conception of a sharply separable dualistic ontology is a hopeless cause.

This is not, however, necessarily a negative conclusion. My goal is not to resolve disagreements across disciplines and over deep theoretical issues; it is to develop a workable method, based on reasonable theory, for evaluating environmental change, a method that will guide us toward a normative approach to sustainability.

3.3. Two Kinds of Rationality

No complex social problems involving multiple interests and values can be comprehensively represented in a system that would allow the computation of a best solution. None of the touted formal models can provide a solution to a problem of conflicting values; this means that a number of approaches to decision making must be rethought. There is no nonprejudicial method of "analysis" that will turn "messes" into tractable "problems."

Some theorists lament this outcome, claiming that the pervasiveness of wicked problems entails that all policy discourse is doomed to conflict and pervasive disagreements. The Western notion of rationality is based on getting to one right answer, whether that answer is extracted from scripture by interpretation or by the march of objective science toward full knowledge. Because the prevalent view of science sees a march toward objective truth—recognizing the vast uncertainties, but treating their elimination as "in-principle-possible"—the myth asserts that more science and more comprehensive models will yield answers. In the meantime, paralysis sets in, as inquirers try to approximate a nonexistent unique path to given objectives.

The sense of malaise that arises when a wicked problem is faced can be escaped if one accepts the limitations implied and explicitly renounces the futile search for uniquely correct solutions. To see uncertainty as a reducible quantity is to deceive oneself. Faith in the IPP is unjustified and continuing to act as if the void can be filled with a little more science is naive. Accepting this truth need not be depressing, however. Once it is acknowledged that all decisions are made under uncertain conditions, the path forward becomes clearer. Just because there are no single right answers, one need not conclude that there are not both better and worse answers.

Now it will be possible to begin to untie the Gordian knot of theory and practice. If ignorance is recognized, if attention is paid to what can be learned when one acts, however uncertain the situation, it will be possible to recognize that learning one's way forward is the only way forward. Neither science nor logic can finally solve the problem of uncertainty. What is needed is a concept of rationality that can help one to make judgments and to learn one's way out of uncertainty and policy gridlock.

Accepting the wickedness of the problems faced, rejecting the false hope of an ideal solution, shifts the action and decision points from theory to practice. Attention must move from calculation to fostering of decision processes that are effective and fair. This shift to a new regime in environmental policy discussion and implementation will be marked by engagement of scientists, the public, and policy makers alike in a learning process.

So, putting the negative connotations of wickedness to one side and embracing the insights of Rittel and Webber, one can see a new direction for research, which will reveal opportunities to improve problem formulation and encourage engagement of affected parties. The age of analysis is being replaced with the age of negotiation. The age of negotiation, in a situation of uncertainty, must in turn become the age of experimentation. Signs of the change can be discovered in the literature, and many social scientists are finding evidence of situations in which newfound faith in negotiation and, especially in ongoing, iterative collaborations, has developed ad hoc to deal with problems perceived in specific communities (Wondolleck and Yaffee, 2000).[3]

What is important to realize is that rationality does not die with the dismissal of the myth of the single right answer; rather, it has to be adapted to the problem at hand, which is to identify a path toward sustainability, even in societies characterized by diversity of values, opinions, and perspectives. Rationality can manifest itself even in the absence of certainty, and the rational thing to do is to act and to act in such a way that our actions function both as first approximations of desired changes and as scientific experiments, complete with controls and the other disciplines of scientific endeavor. Yet these experiments are most useful when they are undertaken within an ongoing collaborative effort involving interested stakeholders who also desire to learn. Learning through action is effective when the interested public takes ownership of experiments and experimental actions. For example, groups of ranchers in the US western states have formed collaboratives to experiment with various grazing patterns to evaluate impacts on vegetation and soil health in grasslands.

Fortunately, while many of today's contributions to understanding sustainability are advanced based on the false hope of a single right answer, an alternative approach to decision analysis of complex problems and how to respond to them rationally is available. Herbert Simon, a polymath and winner of a Nobel Prize in economics for his work on decision analysis of individuals and groups, states the choice clearly by distinguishing two understandings of the concept of rationality: substantive and procedural rationality.

Substantive rationality attempts to identify one correct answer by computation or by some algorithm, and it attempts to determine what is substantively rational by considering and measuring all outcomes against stated objectives. Procedural rationality, by contrast, is determined by the appropriateness of the procedure employed to the problem at hand (Simon, 1976; O'Neill et al., 2008).

Substantive rationality should remind one of Optim, who is searching for sustainability objectives, expecting that a statement of objectives is forthcoming. Substantive rationality has much in common with means-end-rationality typical of scientific resource managers. Actions seeking to achieve substantive rationality, like the specialized dog breeder in section 1.2, define a clear-cut, dominant objective with actions judged according to their success in achieving the defined goal (the largest dog possible—Great Dane—or a "small, cute dog").

Simon provides an alternative way forward; he shows that one can abandon the search for substantively rational responses to wicked problems and enter an alternative world of process studies. These studies of how and why communities sometimes end up in stalemate, and at other times, forge coalitions that can respond creatively and effectively to some aspects of wicked problems, provide an alternative to hopeless searches for an ideal substantive outcome (Wondolleck and Yaffee, 2000). Here, the goal is to approximate an ideal process, one in which all parties are included, pluralism of objectives are assumed, and the like; and good decisions will be ones that have emerged from a process that approaches the ideal. A huge advantage of this procedural approach is that, while comprehensive, substantive solutions to wicked problems seem impossible, it is possible to study empirically which procedures and which heuristics lead to more cooperative behavior, reduction in conflicts, and trust-building. Shifting to a process approach, then, replaces the void faced by advocates of algorithmic solutions with empirical study of what works and what doesn't work in public policy processes. We are now able to introduce Adapt's first heuristic.

Heuristic 1
The process rationality heuristic. Pay attention to process. The rationality of decisions depends more on finding an appropriate process than on making accurate calculations and predictions.

Thus a more robust response than depression to the result of Rittel and Webber is to direct attention to processes by which communities define problems and to provide heuristics to guide such communities. A shift of emphasis from pursuing calculations that cannot be completed to providing heuristics can at least ensure that certain key questions are raised within a public process. By shifting from an emphasis on computation to an emphasis on public processes, a forum can be created that will allow discussion as well as iterative treatment of problem formulation (National Research Council, 1996).

The appropriate reaction to wicked problems is to accept that they have no comprehensive solution and to direct attention to systematic, effective ways of dealing with those aspects of wicked problems that make them wicked. Here, I think the most profound question in public policy analysis is faced: Should one continue to press forward with formal (substantive) analyses such as CBA and risk analysis, trying to push these methods as far as possible, and leave the wicked aspects of the problem that remain within an unexamined black box? Or, should one shift to a new understanding of public decision making in which, rather than pursuing a singular outcome, one faces wicked problems by emphasizing public involvement and public debate in a real community as the main context for decision making? The latter approach encourages the development of institutions that are capable of addressing problems through public discussion; information from the sciences, including economics, in this approach, become contributors to the discussion without being viewed as partial accountings of a comprehensive solution. While addressing wicked problems through iterative discourse and debate, it is, of course, possible to bring more formal analyses to bear within the debate. If one focuses on why some problems are nonbenign and begins to struggle toward a method that is consistent with the Rittel and Webber argument, formal methods can still be used, but they should be seen as serving a larger public process and discourse. This organization of public discourse with scientific models as learning aids is developed in Norton (2005, especially chapter 11 and below in section 3.5).

On this understanding, recognition of the analytic "void" defined above can instead be seen as the transition stage initiated by the creative destruction of the myth of possibly complete knowledge, which will be followed by a shift toward public involvement in adaptive management processes. Adaptive thinking uses evidence carefully and designs interventions as experiments capable of reducing uncertainty within the process. In this view, decision analysis does not solve problems when values conflict—instead, it assists (provides heuristics) for considering possibilities, and it relies on developing appropriate public processes to achieve partial or temporary resolutions.

3.4. Expect Surprises! Introducing Adaptive Management

Adaptive management is the source of most of the wisdom that can be attributed to our character Adapt. So far, the story has been largely negative—clearing the conceptual brush that obscures the role of values in environmental decisions—but now it is time to start building up the alternative, procedural approach to pursuing rational environmental policy. Again, it turns out that what is needed by way of methodological ideas and direction need not be developed from scratch; good ideas and actions are already out there, and those ideas lie on a fairly direct intellectual path from Henry David Thoreau, to the American pragmatists (especially John Dewey), through Leopold and Herbert Simon, to today's evolving conception of adaptive management as understood by Holling, Gunderson, Lee, and the Resilience Alliance.[4]

Development of a public discourse around a problem is a first step in creating institutions and organizations that are appropriate ones to address perceived problems within a community. As Dewey recognized, this first step will lead to further steps if deliberative discourse incorporates the scientific method of experimentation, opening the possibility of social learning (Lee, 1994).[5] In environmental policy development, Dewey's understanding of social progress can form the intellectual and scientific foundations of the "adaptive management" approach, which has several aspects:

1. Adaptive management is experimental management; adaptive managers steadfastly try to apply the scientific method, even in confusing and complex problem situations. This is deemed the most effective response to the unavoidable uncertainties afflicting most environmental problems.

2. Adaptive management views problems from a place. This aspect does not mean adaptive management is always local; adaptive management is multiscalar. But the adaptive manager always views the system of interest, no matter how large, from some geographical place within a complex, surrounding environment. A "place" is understood as a physical site where nature and culture have intertwined to create a unique system.[6]

3. Adaptive management is a problem-oriented approach, and this problem orientation is more basic to adaptive thinking than are organizations and institutions. This means that successful development of an adaptive management process will often transcend existing agencies, offices, and organizations. Building one or more organizations that can bring together members of existing institutions, as responses to a problem that is articulated in public discourse, provides the most effective way to address complex problems, including wicked ones.

4. Adaptive management does not see environmental problems as presenting once-and-for-all decision endpoints toward which an environmental policy is directed. Adaptive management is iterative and ongoing. No matter how long a problem has been addressed within an adaptive management process, it is always possible to revisit the formulation of the problem.

Adaptive managers such as C. S. Holling and Lance Gunderson emphasize that a scientific approach to environmental management, striving to manage for sustainable development, cannot be aimed at a particular result, at a target, or at a set of objectives stated at the outset of a process (Gunderson and Holling, 2001). Rather, even the objectives of management must be evaluated within an ongoing process. The point of adaptive management is to follow the scientific method—the method of experience—wherever it leads. Following that process will result in real surprises that catch one off guard but can reveal new opportunities for experimentation.

3.5. Adaptive Management as Embedded Science

The emergent approach embodies many of the earlier ideas of Aldo Leopold. This approach was first called "adaptive management" and developed in the 1970s and the following two decades by resource-oriented scientists who developed sophisticated mathematical models to predict the future of fisheries and forests, given various levels and patterns of human exploitation. While this mathematically and theoretically sophisticated work continues among many adaptive managers, the idea of adaptive management has developed into a far more comprehensive approach to thinking through management problems. The development of the movement can be charted in three important steps.

The first step in this broadening was a recognition in the 1980s that, given the cumulative impacts of human societies and their economies on natural systems, management modeling must be expanded to include attempts to manage systems that mix humans and natural processes. Before then, most ecologists studied pristine systems, considering human activities as confounding variables. Adaptive managers realized the importance of coupled human and natural systems and included the social sciences more directly in their discussions of management processes and goals. This shift encouraged a coalition with ecological economists.

As a second broadening step, adaptive managers moved beyond modeling of particular variables affected by direct takings from nature (such as fishery yields and productive forests) to model the more generalized effects of humans on the long-term functioning of ecological systems. This expansion of concern indicated the general importance of maintaining the resilience of ecosystems. Since resilience is a system-level characteristic, study of it required shifting attention from counting populations to monitoring system processes.

There is currently emerging a third expansion of the purview of adaptive managers: a recognition of the import of social values for management. As adaptive managers developed their models of human impacts on ecological systems, they realized that, while most economic reasoning emphasizes the values accrued from a flow of natural resources useful in human products and services, their more complex system models also implied that a value should be placed on the underlying processes of nature that indirectly produce the resources consumed in human economies. They learned, to put the point simply, that a resilient system can be of great value to humans (even though resilience itself apparently cannot be directly used or traded in markets), because it creates the conditions for ongoing productivity. The value of resilience manifests not as a single resource among others but as a structural feature of the system that has allowed the development of thriving economies that extract one or more economically useful outputs from the system. Recognizing that loss of resilience can result from ongoing overexploitation, adaptive managers are concerned about the social impacts of "flips" into a new functional regime that can occur as resilience is gradually eroded.

Adaptive managers have progressively become more aware that successful collaborative management requires more careful treatment of social values than was traditionally thought to be appropriate for empirical scientists. In this book, I advocate embedding evaluation of environmental change, and the development of methods by which to evaluate change, within the broadened process of adaptive management. I also advocate embedding these discussions within a social, collaborative framework.

One important outcome of these trends toward a more general approach to adaptive management is the realization that, in addition to facing uncertainty about the structure and functioning of systems, there is also uncertainty about the goal of sustainability and what it will involve. This recognition represents the keystone of a new approach to both environmental values and the search for sustainability. It is the same insight that is signaled in our advocacy of normative sustainability: the recognition that human, environmental values must be at the heart of any discussion of how to live sustainably.

This broadened view of adaptive management only heightens the contrast between adaptive management and SRM. Historically, environmental management was considered a science; further, despite occasional protests to the contrary, it was argued that it could be a value-neutral science. Forest economics could provide an economic model based on maximizing production, so foresters simply planted and cut according to their scientific understanding of how to maximize timber growth. The value they served—economic prosperity—was a political mandate: their job was the technical job of maximizing harvest.

Rejecting this traditional understanding of science as value neutral, it is necessary to focus on social learning in response to real-world problems and applications of the scientific method in all aspects of management and policy; these are essential features of the emerging approach. With this broadened conception of adaptive management, scientifically supported reasoning is brought to bear on values and goals as well as other uncertainties. It involves, more specifically, the Deweyan idea that reason and logic include the entire range of tools afforded humans by language and intelligence, and that there is only one method—that of experience in both science and ethics.

With the establishment of a science of adaptive management, it is important to note that adaptive managers are also steeped in value judgments. To understand the role of science in the science-policy interface, we don't need to take on the general issue of whether science can ever be value free. Rather, it is important to focus on the everyday practice of environmental scientists and environmental managers. However value-neutral science can be under the best of circumstances (in peer-reviewed disciplinary journals, for example), science absolutely is not value neutral when it is inserted into complex negotiations regarding policies. Setting aside the question of the overall value neutrality of science, one can focus on the way that science functions in real situations where actual management problems are being addressed, and where stakeholders have both differing models of the problem and differing interests and values. The situation is characterized by uncertainty—about how to bound systems, how the systems work, and which managerial goals should be pursued.

If adaptive management is to be considered a science—as I believe it is—then it must be a science in a new sense of the word, a sense that draws intellectual boundaries very differently than the positivists did. The most useful discussions of this relatively new vision, or new type of science, distinguish curiosity-driven science from mission-oriented science (Funtowicz and Ravetz, 1990, 1993; Scott et al., 1994). Curiosity-driven science is traditional, disciplinary science—science that is done to advance knowledge and is generated within disciplines that exercise quality control through peer review and other forces. Clearly, adaptive management is explicitly mission oriented; it engages nature through experiments that are designed to increase knowledge that may help us make better choices in the future.

While there is interest in reducing uncertainty, some participants recognize that some scientific findings might undermine their policy positions. As an example of mission-oriented science, we can take scientists who are studying the Everglades, a natural treasure severely degraded by unfortunate past managerial decisions. Scientists in this context often face phenomena to which they have strong normative reactions; many of the scientists have their own views of what is best to restore, while other scientists may be working for a developer or one of the sugar companies that grow sugar and pollute water south of Lake Okeechobee. Even among committed environmentalists and restorers, there are rifts, as representatives of the US Park Service often object to policies pursued by the state of Florida and vice versa. In situations such as this, science functions quite differently: in contexts like this, the scientific method is still the touchstone, though it can be very difficult to apply its logic in a straightforward manner because of the complexity and wickedness of the problems faced.

3.6. An Epistemology for Adaptive Management

Epistemology is the study of knowledge, its nature, and the methods by which humans gain knowledge. Much of the history of philosophical epistemology has focused tightly on the search for certainty. It was widely assumed throughout the history of philosophy that knowledge of anything required certainty about some things. It is clear that such an epistemological approach—wedded as it is to a priori knowledge and to failed claims of certainty—is inappropriate for adaptive management, which assumes uncertainty as an unavoidable aspect of every situation requiring action.

Fortunately, there is an epistemological tradition that is better suited to contexts of great uncertainty and controversies—the epistemological approach of American pragmatists. This epistemology is based on experience and involves an application of the scientific method, as adaptive managers respond to uncertainty with a spirit of experimentalism that encourages learning by doing.

To see how such an epistemology would work, I explain a useful analogy for understanding the progression of human knowledge due to the German sociologist Otto Neurath. He suggests that one should think of our body of knowledge as similar to a seafaring ship that goes from port to port but that cannot go into dry dock for repairs, for various reasons (Neurath, 1932/1983). Engineers and craftspeople must thus repair the boat while en route—a tricky but not impossible task. Neurath then asks one to imagine that this process continues indefinitely and that, over time, as the ship's planks and rigging wear and rot, all the planks and other elements are replaced one by one or in modules, while keeping the ship afloat. Over time, perhaps every element of the ship will have been replaced, completely renewing it without changing its identity or its function.

To apply the analogy, Neurath argues that, similarly, there are no intellectual dry docks where one can step out of the current framework of beliefs and assess it as a whole, independent of all assumptions and beliefs. One must, rather, continue to use language as well as our current concepts and beliefs to communicate, even as corrections are made in some areas of our belief system. This clever analogy captures several key themes of a pragmatist understanding of knowledge development. Five such themes have been listed by Richard Bernstein (1997), who summarizes pragmatist epistemology in this fashion:

1. Antifoundationalism, which rejects the view that knowledge must ultimately rest on a foundation of noninferential, self-justified beliefs. Pragmatists do not believe that a priori, or necessary, propositions are required to support other beliefs The early pragmatist Charles Peirce suggested that knowledge is supported by other knowledge that has the same contingent status, and he argued that one should not think of reasoning as a chain of inferences but rather as a cable that gains strength by the intertwining of many threads.

2. Fallibilism, which asserts that every belief must ultimately be examined and verified. No assertions can be established except through empirical study.

3. Radical contingency, which is the belief that the universe is wholly contingent and that chance is an inherent aspect of the universe and human knowledge of it. Neurath's boat analogy captures the first three Bernstein themes: every belief should be considered vulnerable to refutation by new experience—and every plank will eventually be examined by the method of observation. Nevertheless, the ship of knowledge can stay afloat as inquirers stand on so-far-unrefuted beliefs while other beliefs are examined, just as repairers of the boat stand on the strongest planks to upgrade weaker ones, recognizing that they may in the future examine the very plank they stand on today.

4. The social character of the self and knowledge, which rejects the idea of individualism with respect to gaining and justifying knowledge. Objective knowledge is not considered the domain of the individual mind: knowledge production and verification is a social endeavor, requiring many viewpoints.

5. Pluralism, which can be understood as a description of the actual situation: as one looks at individuals and groups, one finds advocacy for multiple, sometimes inconsistent, beliefs. Unlike traditional philosophers who believe that, when different and apparently conflicting beliefs clash, one or the other must be wrong, pragmatists believe that there are multiple frameworks of knowledge and that humans can benefit from a mix of differing viewpoints. The examination of weaker planks on Neurath's boat will be more successful if it is critiqued by individuals with multiple areas of expertise: an examination by an expert woodworker, by an engineer, and by a boat-builder will be superior to efforts by just one of these.

I suggest that Neurath's nifty analogy, then, can provide a quite good outline of an epistemology for adaptive management: in the face of uncertainty, one should first doubt everything, but not all at once; since there are no a priori, necessary truths, one should next expect that the collective wisdom of a diverse group is more likely to contain better ideas than those of a closed society of experts; and finally, one should encourage discussion and debate among representative groups of a public as a means to reach agreement as to what to do. Explaining this general viewpoint, Bernstein quotes Richard Rorty's take on the pragmatic approach: Rorty says that what makes attempts to arrive at a final truth so difficult "is the growing realization that there are no uncontested rules or procedures which will tell us how rational agreement can be reached on what would settle the issue on every point where statements seem to conflict." Bernstein concludes, "Here one begins with the assumption that the other has something to say to us and to contribute to our understanding" (1997, p. 399). We are now able to introduce Adapt's second heuristic.

Heuristic 2

The epistemological heuristic. Trust experience more than ideology; question everything, but not all at once. Be experimental. Robust approximations are better than precise projections.

Does this epistemology entail relativism or nonrationalism? It does not. In fact, it supports an attractive position between dogmatic certainty and abject skepticism, a position I will call (for lack of a better name) "limited realism." I label this position "realism" because it features a belief that the sensory input that one encounters is not of our design, but rather one is bombarded by sensory input that we do not control. It is real and objective in the specific sense that it is independent of our will and choice. This form of realism is also quite "limited" in an important sense, as the ways that one sorts through processes and organizes the flow of independent information vary in highly individual manners. One cannot assume that because two individuals are in the presence of the same observable phenomenon they will agree about what they observe. This is the old nemesis, the factor that contributes to the wickedness of problems: different individuals, with different life plans, different projects, and different concerns, tend to see situations, messes, and wicked problems through such different lenses that full communication is impossible and cooperation is unimaginable.

Pragmatist epistemology, however, provides a more nuanced approach to wicked problems and those epistemological problems caused by both uncertainty and conflict over interests and values. Following Neurath, one must put the repair workers (by analogy, the scientific practitioners and the evaluators) on the ship (by analogy, they must be engaged within the processes that are unfolding locally, in a place, with particular physical, political, and social features, and they must communicate and act within the discourse of the community or affected parties). And the scientists or repair workers must work continuously and iteratively, making unavoidable assumptions today in order to better study a particular area, but then reevaluating those very assumptions tomorrow. Limited realism, then, is the method of repeatedly applying the scientific method in the face of bombardment by new and sometimes novel stimuli, making assumptions, and trying out the assumptions as various ways of understanding our sensory inputs. If participants, in an attempt to manage their place, undertake an iterative process, seek partial solutions, build trust among competing factions, and place their faith in the scientific method, progress can be made. This progress, which will be measured in greater cooperation and the further development of effective managerial solutions, is the promise of adaptive management.

The epistemology here proposed is not powerful enough to falsify propositions finally, nor can it yield certain conclusions about the world from science or logic. This epistemological method, then, promises only that, given enough time and sufficient exchange of ideas, beliefs are corrigible. Worse ideas will be set aside in favor of those that better help inquirers to make sense of common experiences. Participants who seek cooperation need not come from the same perspective or agree on everything as long as they try in good faith to understand each other on the way to cooperation. This epistemology is enough for the adaptive management process to struggle toward both cooperation and a more widely shared view as to what sustainability is.

Adaptive management is therefore the central framework here suggested for building better environmental policies and ultimately sustainability. Adaptive managers, while cognizant of vast uncertainties facing communities with complex, often wicked problems, can put their faith in limited realism and the human struggle to characterize, describe, and manage their interface with an outside world that evolves independent of human effort. One often faces confusing information, yet the way to deal with it is to develop appropriate language and to engage in communication with associates, many of whom will have ideas and perspectives unlike ours. Such activities will not lead to an ideal solution; such activities may, however, when pursued diligently and iteratively, result in practices and activities appropriate to the context at hand. The goal, then, is not to arrive at a perfect truth but to create a process that is corrigible through the sharing of experience and the exchange of ideas.

CHAPTER 4

Contesting Sustainability: Who Will Own the Word?

4.1. What Is Sustainability? And, What Is to Be Sustained?

Sustainability is a contested concept. While it unquestionably refers to a moral relationship across generations, theorists and practitioners of different disciplines differ in the way they characterize and evaluate this relationship. There is a vast literature on the meaning and practice of sustainability, both in technical fields and in the general literature of environmental literature.

As was explained in section 1.2, sustainability gained prominence in the 1970s and, especially, in the 1980s with conferences and publications, especially the much-discussed report from the World Commission on Environment and Development, *Our Common Future* (World Commission on Environment and Development, 1987). The report defined sustainable development as "development that meets the needs of the present without compromising the ability of future generations to meet their own needs." This definition has stimulated much use and considerable discussion because, for one thing, it characterizes sustainable development mainly in terms of availability of resources to humans without direct reference to, for example, ecological limits.

Different approaches to sustainability and sustainable development have advocates among the practitioners of several disciplines, each of whom adapts the concept of sustainability to their own theories, concepts, and models. In particular, the social scientific field of economics and the natural scientific field of ecology have attempted to articulate sustainability principles that embody the concepts and wisdom of their fields. Indeed, once authors of these two fields had offered their insights on the topic of sustainability, it became obvious that they were offering "models" that differed in their most basic assumptions and implications (Norton and Toman, 1997). Faced with the apparent conflicts that these insights seemed to imply, a group of interested and creative scholars from those two disciplines, as well as interested parties from related fields, founded a new field, ecological economics, which is sometimes referred to as "the science of sustainability" (Costanza, 1991; Clark and Dickson, 2003).

As it turns out, the differing insights of economics and ecology encouraged the development of different conceptions of sustainability and set the stage for ongoing contestation (Beckerman, 1994, 1995; Daly, 1995). Economists employed market models while ecologists generally proposed resilience models. Economists measure change as variation from a set point: they represent the value of objects and processes as their price in exchange. Price changes are then measured within partial equilibrium systems, with the expectation that such changes will explain deviations from equilibrium and also reversion toward the prior steady state. This value, reflected in prices measured as willingness-to-pay (WTP), in turn reflects the preferences of consumers. As such, the models represent behavior of human individuals within the complex social institution called a market. These models purport, then, to represent a description of the value that consumers (the subjects of the analysis) place on objects by recognizing that they will choose differently as prices of commodities vary, so individual choices are expected to be informed by knowledge of future scarcities, and so on. While such models represent change, they also embody an expectation that, as conditions and choices change, the behavior of the partial equilibrium system will fluctuate around an original fixed point as the behavior of individuals varies in response to economic conditions.

Resilience models, by contrast, focus not on human behavior but on the condition of systems that provide the context and resources for human activities. Ecologists model a system of human behavior, one in which human activities are only a single component. Resilience modelers consider the possibility of multiple "basins of attraction," and when ecologists represent the dynamics that concern them—the changes that seem to them to threaten sustainability—they pay attention mostly to "flips," or discontinuities in the functioning of ecological systems.

These differences can be illustrated by reference to a muffin tin (a baking container with multiple small indentations designed to produce individual cakes). If one places a marble in one of the indentations, one can shake the tin and, if shaking is severe, the marble will fly into another indentation. Equilibrium systems, such as those studied by economists, can be modeled

Figure 4.1. Resilience in a muffin tin: A muffin tin—a baking dish with multiple indentations— illustrates two kinds of resilience. For economists: a marble displaced up the side of a single indentation by a disturbance will return to equilibrium, representing resilience as understood by economics. For adaptive managers: a marble displaced by a stronger disturbance into another indentation has entered a new regime, signaling lack of resilience. The system is sometimes said to have "flipped" into a new basin of attraction.

within one indentation—changes in the system stay within a single basin of attraction and, when disturbances to the system occur, it is expected that, when the shaking stops, the system will return to its original location in the bottom of the indentation in which it was originally put. Economic equilibrium models can be represented by one indentation, while ecologists, who fear flips into another regime, worry that sufficiently violent changes will result in the system's flipping into a new equilibrium (ending up in a different indentation).

Practically, flips often represent a threat to natural resource producing systems and can lead to economic tragedies such as when a fishery collapses, causing bankruptcy of fishers and devastation to local economies. Resilience models, as illustrated by the muffin tin in figure 4.1, can measure change from set points, but expect that severe or prolonged stresses can cause a system to shift ("flip") into a different functional regime, creating a new equilibrium. If that new regime provides a less favorable background system for the pursuit of human activities, a failure of sustainability has occurred.

These points about modeling by ecologists and economists are important because, when practitioners of these fields turn their attention to sustainability problems, they do so through the lenses of their disciplines and therefore favor system models that have been effective for them in addressing problems in the past. Central to this controversy is a conflict between economists and ecologists on a related question: can resources that become scarce be replaced with substitutes? If one believes that all resources have an adequate (economic) substitute, then one will be comfortable with a system of analysis that rules out irreversible flips—an equilibrium system. But if one worries that some losses of resources and some damage to ecological systems will result in irreversible change, one will prefer the muffin-tin model.

It turns out that, if one pays attention to this difference, one can place theorists on an intellectual "geography" of sustainability conceptions. This geography, represented in table 4.1, creates a useful categorization of the sustainability conceptions (and the associated "models") of proponents of various answers to the two questions. Columns in the table represent differing conceptions of sustainability, and in the rows, one sees the variety of definitions and conceptions that have been introduced by commentators with different models, including economics, ecology, systems theory, environmental philosophy, public policy, and the like. Rows describe positions that are determined by one's understanding of equilibrium versus nonequilibrium systems and thereby indicate which "paradigm" the answers are associated with, mentioning prominent advocates of each position. This classification system is intended to map the intellectual territory between "absurdly weak sustainability" (which is off the chart to the left) and "absurdly strong sustainability" (which leaves present humans with a diminished right even to live and reproduce and which falls off the chart to the right). The two "absurd" positions represent limiting cases on either side—positions not held by anyone—and the chart taxonomizes positions falling between them. These four positions have been articulated and defended and will provide useful prototypes that can be compared, criticized, and evaluated for their usefulness. Here, I briefly explain how each of the four positions reflect disciplinary perspectives and assumptions.

4.1.1. Weak Sustainability

Column 1 represents "weak sustainability," which is popular with mainstream economists. According to this view, which takes its models and insights from mainstream, neoclassical economic growth theory, sustainability is assured as long as savings in each generation are sufficient to avoid de-

Table 4.1. A conceptual geography of sustainability definitions

	Weak sustainability	(Strong) economic sustainability	Strong sustainability	Normative sustainability
Home discipline	Mainstream economics	Ecological economics	Systems ecology	Policy science / environmental ethics
Paradigm	Welfare economics	Welfare economics + natural capital designations	Complex dynamic system theory / adaptive management	Complex dynamic systems theory / adaptive management
Definition	Maintenance of undifferentiated capital	Weak sustainability + maintenance of natural capital	Weak sustainability + maintenance of resiliency	Weak sustainability + maintenance of options
Key concepts	Nondeclining wealth	Maintaining natural capital	Maintaining resilient ecosystem	Integrity of place / community involvement
Key advocates	Solow	Pierce and Barbier	Holling, Lee	Leopold, Norton

Weak sustainability–Welfare counters ⟷

Strong sustainability–Stuff counters ⟷

Source: Adapted from Norton (2005), p. 314.

clines in the general capital (economic wealth) of the society. According to this approach, our moral obligations to the future are fulfilled if people of the future have the opportunity (stored in capital) to develop efficient technologies and achieve levels of welfare at least equal to ours. In other words, weak sustainability theorists believe that it is the economic system of exchange and prices that should be consulted as the key system dynamic. Further, the analysis of that system measures only one variable—general economic wealth—and sustainability must, on these assumptions, be represented as the path of economic growth from the set point of the current economy. Change affecting sustainability will be measured as deviations from the given set point, which typically is taken as the level of current wealth, measured as total, undifferentiated capital, including both human-built capital and capital stored in nature. A society is sustainable, in this approach, if that society has a savings rate sufficient to ensure that no decline occurs in this measure of overall wealth (Solow, 1993).

As will be discussed further in section 4.2, the keystone assumption of weak sustainability as represented in column 1 is that, in an economic system based on markets and on exchange value, consumer choices represent value. This assumption implies that if the price of an object increases by too much, consumers will accept a substitute plus the savings from avoiding the more costly item. So the assumption, in turn, supports the view that, for any resource of value, there is some substitute for it; and the assumption of universal substitutability seems to call into question the ability to specify any particular feature of the world that must be preserved if harming the future is to be avoided. If people of the future are wealthy enough to reward innovation, their economy will generate substitutes for any scarce resources. As increasing scarcity of those resources drives market prices higher, entrepreneurs will be incentivized to develop technologically based alternatives. Column 1 thus represents an approach to sustainability that focuses mainly on sustaining the economic opportunities available in successive generations of a society. Sustainability is achieved by making wise decisions about investment and consumption, thus maintaining at least undiminished capital.

4.1.2. Stronger Economic Sustainability
The second column represents a valiant attempt by some economists and ecologists to find a more integrated approach to sustainability that remains economic in its evaluative function but that takes some types of ecological change to be damaging to the future. According to this group of ecological economists, welfare economics and the usual economic models that measure it are only one part of the story. Earlier generations not only needed to protect

the store of total wealth; it is argued that they must also have protected crucial aspects and processes of natural systems, which they characterized as "natural capital." If we, at first, assume there is a clear definition of natural capital (natural features that have no artificial substitutes), this brand of sustainability would require a nondeclining stock of both kinds of capital. This apparently attractive position, while perhaps useful and interesting as a sort of placeholder for a possibly emergent integration of ecology and economics, has not yet provided a clear and distinct alternative to weak sustainability, because its advocates have not been able to define natural capital in noneconomic terms.

Indeed, the label itself—the phrase "natural capital"—seems prejudicial in that "capital" introduces an economic analogy. If one defines natural capital as an element or process of the natural world that, if lost, will reduce the economic welfare of people in future generations, natural capital, in this case, is defined in economic terms, reducing the idea of natural capital to its effects on human welfare. This view seems, ultimately, not to differ from the weak sustainability approach of column 1, since ordinary weak sustainability will include all welfare effects and calculations. If, on the other hand, "natural capital" is defined in terms that are independent of effects on the measurable welfare of future generations, one is faced with a conundrum: how will the value of natural capital be represented in "natural" terms (i.e., noneconomic, nonwelfare impacts), and how will this noneconomic value be compared with and count within an economic system of valuation? Unless strong economic sustainability theorists offer a noneconomic criterion for identifying elements of nature that constitute "natural" capital, this approach is best understood as a variant on weak sustainability, which emphasizes natural endowments in the pursuit of economic sustainability measured as welfare, but which fails to offer a concept distinct from measurement of opportunities for achieving economic welfare.

4.1.3. Strong Ecological Sustainability

There is a group of scholars, some of whom are mentioned above, who function together in more or less formal ways with ecological economists, a group that is the contemporary embodiment of the adaptive management movement. This group, sometimes referred to as the "Resilience Alliance," has expanded its purview beyond modeling of natural systems and has developed the Darwinian idea of adaptation as an analogical approach to understanding complex obligations beyond obligations to maintain wealth. Adaptive management represents a leading example of "strong (ecological) sustainability" and includes many members of the field of ecological economics and members of the Resilience Alliance. According to the conceptual geography, adaptive managers (and other strong sustainability advocates) understand sustainability in the context of multiscaled, complex-dynamic systems that can equilibrate at multiple points. So they see the descriptive concept of resilience as important to management goals. This approach, implicitly at least, recognizes that "resilience," when stated as a goal of ecologically sensitive management, embodies a value. Resilience functions as a hybrid, a concept that (one hopes) carries a descriptive measure and also embodies a value worth managing to protect.

For conceptual completeness, I define a position, as captured in column 3, as advocating a "purely" scientific concept based on the measurable concept of resilience of systems. This "pure" scientific-descriptive approach would combine the use of complex, dynamic systems theory and would emphasize management for resilience of systems on multiple levels, while insisting the approach is purely empirical, turning sustainability into a scientific concept (sustainability simply becomes the measurable resilience of ecological systems). A number of theorists in this area refer to their work on sustainability as "sustainability science," and while emphasizing that the scientific side of sustainability need not conflict with recognizing the importance of values in defining the concept, placing science in such a dominant role weakens these scholars' contribution to the important value problems faced (Costanza, 1991; Clark and Dickson, 2003).

For reasons explained in the previous paragraph, it may be necessary to see this position as a straw man, with nobody really arguing seriously that the study of sustainability can be entirely based on value-neutral science, as all strong sustainability theorists have progressed significantly across what turns out to be a very porous divide between strong sustainability theorists (column 3) and what I have called "normative sustainability" (column 4). In the process, they have virtually erased any sharp line between strong ecological sustainability theorists and normative sustainability theorists. The more one emphasizes the importance of social values and their relationship to physically defined features of natural systems—such as resilience—in the search for sustainability, one has embraced "normative sustainability," even if those values are associated with a descriptive concept of resilience.

4.1.4. Normative Sustainability

If one looks at the sustainability challenge from a broad perspective, one must understand "resilience" as a descriptive characteristic that also functions as a value. The idea of resilience, which involves the inherent ability of a physical system to respond to disturbances and return to its previous dynamic trajectory ("homeorhesis"), evokes a medical analogy with health and integrity. In medicine, the commitment to protect and improve health places an unquestioned value at the heart of an enterprise in which science, experimentation, and descriptive studies predominate. When resilience (or any other system-level characteristic) is understood as a descriptive variable that also tracks a characteristic valuable to human beings, strong sustainability takes on a normative aspect (Walker and Salt, 2006). Resilience is important to support many other values, from natural ones (habitat for wild species) to more socially conditioned values such as sense of place.

Ecologically informed strong sustainability theorists would agree that their descriptions of resilience can guide management goals only when ecology is embodied within management, and only when goals and values come to bear on decisions affecting larger systems and future sustainability. While embracing the work of ecological modelers, one must, to use it in articulating truly strong sustainability, expand the focus from the function of models in describing resilient systems to the more normative task of learning how resilient systems can be encouraged through sensitive management. Once one accepts that social values are an unavoidable element of any search for sustainability goals, and once these values are injected into an adaptive management process, we focus attention on choosing models that allow measurement of resilience and associated characteristics but that also help us to figure out how to manage resilience in ways that are sustainable on multiple scales.

I thus return to Aldo Leopold's wisdom, which incorporated ecological and economic wisdom in a simile: "thinking like a mountain." Leopold recognized that humans value nature in multiple ways, ways associated not just with one but with many natural dynamics, always changing in the face of multiple drivers. Learning to think in the long term, what Stuart Brand calls "the long present," in which many human actions and policies play out over multiple scales, involves paying attention to system scales that we usually ignore (Brand, 2010). Leopold and Brand encourage the same perspective and motivation, the perspective from which one sees and takes into account all the predictable impacts of one's actions on both the near term and the future.

Once one commits to a normative conception of sustainability, then one must have some means of assessing gains and losses in value across multiple scales as they affect multiple generations. Assessing paths into the future requires projecting the values, aspirations, and ideals of present culture into the future, and an accurate evaluation of those paths will require multiscaled and multicriteria tests, in order to accommodate pluralism in environmental values. Any multicriteria test, following Leopold, will also be multiscaled. I believe that recognizing multiple dynamics and multiple scales, and paying attention to impacts on those multiple scales, is the key to finding an integration of the models of both economists and ecologists. It is thus possible to develop an approach to sustainability that focuses attention on the assessment of management goals according to their ability to maintain ecological options that are important to maintain the core values of the community.

One cannot know what to sustain until one knows what has value; nor can one learn what is sustainable for a community by applying a cookie-cutter approach to local situations. To determine what is of value and must be saved requires deliberation and negotiation, and good policies will be ones that serve as many social values as possible. Normative sustainability must thus involve a careful exploration of the value commitments of a community and a concerted effort to determine which natural, physical dynamics are productive of those values and which support the opportunity to exercise the actions and behaviors associated with those values.

"Normative sustainability" is the label I have given to my version of what sustainability does—and should—mean. This concept incorporates the approach to modeling undertaken by advocates of "strong (ecological) sustainability" (column 3 of table 4.1), such as adaptive managers. It is useful, however, to represent it in a separate column from number 4 in order to make more explicit a point that adaptive managers from ecology and resource management have left largely implicit: certain ecological states, such as resilience and diversity of ecosystems, are value-laden terms. Adaptive managers should agree with the point that a full-blown local conception of sustainability must articulate goals broader than economic values. Locating normative sustainability as a separate column merely makes explicit that a sustainability concept that places value on ecological processes has progressed beyond the boundaries of purely descriptive ecology.

Normative sustainability must be an expression of values that are discoverable in the present. (Otherwise, it would offer no guidance for today's activities.) This may suggest that the values that will lead a community to sustainability are simply the current values of the present people who live there. On the one hand, this suggestion expresses a profound truth: if one is to articulate a set of values that will lead, democratically, toward sustainability, they must be values that emerge from and are specifiable in the present. Since living persons are ignorant of the actual preferences of the future, they must pursue the community's relevant values.

On the other hand, current preferences of members of the population of most developed countries—and less-developed ones as well—are directed at the short term and involve desires for consumption levels that, if acted on consistently into the distant future, are unlikely to lead to global sustainability. Accordingly, when one talks about the values of a community and how these must be articulated and protected on the way to sustainable living, one is referring to those community values in an emergent sense—values a community would protect, provided they emerge from an appropriate process. If community members currently pursue short-sighted, materialistic values that drive unsustainable lifestyles, normative sustainability must incorporate appropriate processes by which to reconsider and revise those values. The pursuit of normative sustainability thus does not pursue the indefinite perpetuation of current values in stasis. Normative sustainability, rather, is defined hypothetically: it sustains the values a community would espouse as the outcome of an appropriate process of deliberation and social learning (Norton, 1987, chapter 10). The fact that preferences and values change over time only makes it more urgent that communities engage in a process that will identify aspirations about the future rather than consumptive demands of the present.

4.2. What Should We Measure? Sustainability in Economics and Ecology

4.2.1. The Progression from Weak to Strong

One can read table 4.1 and its columns as points on a continuum from "weak" to "strong," with strength representing an increasingly stringent and demanding set of sustainability constraints on current activities and behaviors. Absurdly weak sustainability (running off the table to the left) would require only that we simply engage in business-as-usual; column 1 represents the minimal requirement that the savings rate must be carefully calculated and observed so that earlier generations do not impoverish the future. Maintaining an adequate savings rate—one that will cause no decline in wealth—supports social values independent of intergenerational obligations, so this position, column 1, is characterized as "weak" in that few restrictions, in addition to reasonable economic principles and a fair savings rate, are imposed to achieve sustainability. Strong economic sustainability—column 2—adds the requirement that "natural capital" be protected. "Natural capital," however, has not yet been given an unequivocal definition, and until one clearly provides a means to separate natural from other capital, this idea does not distinguish it from column 1.

Stronger approaches represented in the last two columns impose constraints on activities that would damage ecological systems, with normative sustainability imposing not only protections for resilience but also protection for specific ecological features that support opportunities that communities consider of high value. The progression across the horizontal space from column 1 to column 4 of our geography begins with protections of economic wealth and adding more and more specific constraints, which are stated in columns 3 and 4, terms not reducible to economics or human welfare. This step represents a decisive move toward a view of sustainability that is based on the protection of physically describable aspects of natural systems, in addition to whatever economic constraints are placed on the actions of earlier generations that express and act on their concern for the future. If one were to carry this move to an extreme, requiring that each generation leave the world as it found it, then that position falls off the right edge of the table. It embraces absurdly strong sustainability, which is not feasible.

While this view of the progression as a continuum provides a useful image, it fails to emphasize a key point that is represented by the heavy vertical line in the middle of the table. A major change in what kinds of values get counted occurs as one moves from column 2 to column 3. In describing this important transition elsewhere, I explained the sharp break in question by referring to a conversation I once had with another philosopher who found the question of what we owe the future as puzzling as I do. After agreeing that the problem is important and difficult, my interlocutor said, a bit exasperated, "Well, we can either count and compare welfare or we can count and compare 'stuff,' but if we are going to talk about fairness across generations, we are going to have to count and compare one or the other over multigenerational time." He was implying that, if one is to meaningfully discuss the fairness of one generation to subsequent ones, there will have to be some kind of a yardstick to measure and compare well-being across generations. Economists and many other utilitarians have clearly opted for measuring and comparing welfare: members of subsequent generations will be as well off as members of earlier ones only if they have an opportunity to achieve levels of welfare no lower than those enjoyed by earlier generations.

Economists, in particular, interpret accumulated and undifferentiated capital as a measure of the ability of an age cohort to attain levels of welfare equal to those achieved by earlier generations. Economists, then, especially ones who work with models of growth and savings rates, are able to interpret sustainability as a comparative measure of welfare opportunities that, in turn, are operationalized and potentially measurable as the measure of total capital—that is, total wealth—of the society at different times.

If one crosses the heavy middle line in our geography, however, and if one begins to specify features of physical systems and processes that must be protected for the future, then sustainability constraints must also be specified in ways that cannot be reduced to the unitary measure of societal wealth. Speaking theoretically, one steps across the line when one identifies some aspect of physical systems—what my friend called "stuff"—as requiring protection, and abandons the economists' commitment to representing all things, both good and bad, in terms of countable impacts on human welfare. For lack of a better term, I adopt, half tongue-in-cheek, my friend's term "stuff" as representing any aspect of the world that is described in physical terms; it therefore contrasts with the more common term for human well-being, "welfare."

This is the most important distinction in understanding the meaning of sustainability and the differing definitions that confuse discourse about our obligations to the future. Definitions to the left of the solid line interpret the relationship between present generations and future generations as a comparison of access to levels of welfare at different times. If subsequent generations, assuming equality of effort on their part, can attain the same level of welfare as prior generations, then weak sustainability has been achieved. At given times, all these theories and associated definitions rest sustainability on a comparison of human welfare. By contrast, definitions to the right of the "welfare-stuff" divide impose requirements to protect stuff—goods, features, and other aspects of physically described systems that must be protected for the future.

Following this reasoning, one can test theories and their associated definitions of "sustainable" with regard to the side of the table on which they fall by asking and answering the following question: is it, on the theory or definition in question, possible that future people might be *richer* (in economic terms) than their predecessors, and still be sufficiently worse off than them to render the judgment that the predecessors did not live sustainably? If, on the theory in question, a generation that is economically richer than their predecessors cannot accuse them of unsustainability, then the theory is a weak sustainability theory. If, on the other hand, it is logically possible on a theory and set of definitions for a subsequent community to convict earlier generations of morally unacceptable unsustainability by citing not welfare deficits but absence of "stuff," even if they have equal or greater wealth, then that theory is a theory of strong sustainability.

To give a very simple example to illustrate how this criterion for separating strong from weak sustainability theories works, suppose that economic growth is sufficient to increase per capita income and maintain a nondeclining stock of capital wealth over generations of a community, but that subsequent generations, as a result of waste and profligate use of water resources, suffer from terrible shortages of fresh water. To state a requirement that would rule out this wasteful behavior, one would require that each generation protect an adequate supply of clean, fresh water. Such an outcome, based on the criterion introduced here, would be unsustainable, according to a strong sustainability theory, but perhaps would be sustainable according to a weak sustainability theory if most people of the future have greater welfare opportunities (and perhaps find solutions to water shortages, such as desalinization).

4.2.2. What Is at Stake?

The table, and the progression from weak to strong that it represents, relates definitions of sustainability as proposed by practitioners of several disciplines. How do these differences matter in defining sustainability? About fifteen years ago, I joined forces with Michael Toman, an economist colleague, and set out both to explore the conceptual and theoretical resources of economics and ecology and to compare the ways in which practitioners of the two disciplines understand, model, and evaluate "sustainable" systems (Norton and Toman, 1997). We recognized that we were in contested intellectual territory, but we optimistically set out to create an "integrated" concept of sustainable systems—one that incorporates key ideas of economists and ecologists into an overarching system capable of both describing and evaluating policies proposed to support sustainability.

We immediately recognized two areas of conflict. First, economists and ecologists disagree on the "accounting problem." Yet these varying approaches differ more radically than most actual accounting disagreements. Ecologists and economists do not merely disagree about the magnitude of certain values, they cannot even agree on what has value and how such value should be measured.

Second, our study of the literature of both fields, and of the models used to talk about long time frames and continuity of conditions across time, drove home to us the point that economists and ecologists deal with continuity and change very differently. We referred to this nest of issues as "the reversibility and substitutability problem." This problem, which lies at the heart of the separation between weak and strong sustainability, will be prominent throughout the remainder of this chapter.

Toman and I learned that there is no smooth intellectual transition to be made between economic and ecological models of sustainability, mainly because the models of the two disciplines are based on conflicting assumptions and different understandings of change over time. Economists tend to assume background stability, and they believe in high degrees of substitutability of one type of resource for others and of human-made objects for natural processes. Irreversibility, for an economist, is a rarity, and it is dealt with as a special case; generally, economists assume that shortages and discontinuities will be corrected by substitutes generated by market forces. The threat of extinction of a species might, for example, be given high policy priority because it is irreversible. But this would be a special case, and economists, relying on the fungibility of resources, do not expect most decisions to threaten irreversible change. While policies chosen will no doubt affect the economic system in multiple ways, these changes are viewed against the backdrop of a partial equilibrium: the system will absorb impacts of choices and reequilibrate within a single basin of attraction.

Ecologists, by contrast, see systems as multiscaled and, while some ecological elements and processes may be redundant, ecologists recognize how difficult it would be to find how much redundancy there is in dynamic systems and where it is located. Thus, for ecologists, losses of elements always represent cause for concern. They see and describe many systems where overexploitation has caused systems to flip into new and less-favored regimes of functioning. This concern is expressed in the muffin-tin analogy: too-violent disturbances above some threshold of intensity may threaten the equilibrium assumed in the economic models. Ecologists worry about outcomes that destabilize normally slow-changing physical systems, while economists operate on models that calculate change at the margins and that cannot represent discontinuities and threshold effects. There is no way to model a flip from one steady state to another in an equilibrium system; the ecologists' worries exhibit a complexity that is unreachable by economic modeling (Holling, 1996).

Ecological economists have thus been skeptical of mainstream economists' understanding of ecological systems as simply a "sector" of the economy. Ecological economists and others have suggested that economic models should be integrated within ecological ones. This is a good suggestion, but actually constructing ecological economic models has proven difficult, for reasons discussed below.

This seemingly unbridgeable theoretical and conceptual gap between ecological and economic models was evident to Toman and me in the mid-nineties, but we also noticed possible integration. He noted that economists, when thinking about normal problems of production and consumption, generally assume substitutability of one resource for another—such as in the substitution of PVC for copper piping in plumbing—and think that the substitution will have no lasting effects on welfare or on economic calculations nor will it destroy equilibrium. He also noted, however, that assuming universal availability of acceptable substitutes seems less appropriate as one examines larger-scale resources and resource-producing systems. For example, an economist might point out that the Great Lakes fisheries have undergone several transformations as a result of industrialization, transportation, and invasions of species from ship ballasts and through the opening of the Saint Lawrence Seaway. These changes wiped out earlier species and replaced them with others. Today, however, fishing remains a major economic contributor to the "lake state" economies, nonetheless, because the introduction of coho salmon has created an economically lucrative sport fishing industry. Despite ecological calamities, substitutes were found for lost species. On the other hand, Toman the economist noted that, if one begins to worry about the health and systemic functioning of the Great Lakes ecosystem, then it is very difficult to imagine a substitute for this large and important system. Examples such as this directed our attention to the following hypothesis: while economic systems are usually effective in providing substitutes for consumable resources that become scarce yet remain in demand, the assumption of universal substitutability is less plausible and useful as attention shifts to larger scales that involve larger systems.

4.2.3. Why Weak Sustainability Is Too Weak

It is not surprising that economists and others find weak sustainability attractive. As noted above, this approach makes the choice to live sustainably relatively painless, since achieving sustainability can be described as a policy that can be justified on economic principles guiding investments promoting economic growth, principles that economists would accept on grounds independent of special concern for the future. This approach is consistent with economists' general preference for minimal intervention in the operation of markets. Weak sustainability, in this approach, describes a path forward that requires nothing more of us than acting according to the guidance of economic rationality and a fair savings rate.

Yet this aspect is only one of the reasons that weak sustainability is so popular among some economists and decision theorists. Another set of reasons for favoring weak sustainability is its apparent promise to offer (at least in principle) a complete and potentially computable approach to specifying sustainable outcomes. Weak sustainability theorists, that is, have proposed a remarkable simplification—I will call it "the grand simplification"—whereby the apparently complicated question of how to live sustainably can be reinterpreted as a question that can be answered by specifying a "fair savings rate." That rate is simply the rate of savings necessary for a given generation of a society to pass on an undiminished stock of economic capital to the subsequent generation. Of course, such a reduction of a complex problem like how to live sustainably to a comparatively simple problem like calculating a fair savings rate is a very attractive outcome. Indeed, it is exactly this simplification that encourages economists to follow Optim in pursuing an optimizing strategy. If, by determining a fair savings rate and using standard economic projections, one can specify today what is a sustainable outcome for a given point in the future, one would have provided a calculable objective that can be associated with sustainability outcomes. This specification, if successful, would provide the narrow objective necessary to proceed toward a substantively rational evaluation of policies proposed to contribute toward sustainability. So it is not difficult to account for the attractiveness of the weak sustainability position to economists and others who seek to justify their sustainability positions as substantively rational.

While it is difficult to take a general stand against simplicity, I follow Albert Einstein, who said, "Everything should be made as simple as possible, but not one bit simpler" (Einstein, n.d.). Sustainability is a more complex problem than is thought, so I believe any grand simplification is an oversimplification and fails the Einsteinian test. If we ask, "What do we owe the future?" we are in fact asking at least four puzzling questions:

1. *The tradeoffs question.* How should an earlier generation balance concern for future generations against concern for their own moral and prudential issues?

2. *The distance question.* To clarify the meaning of "future," it seems important to inquire how far into the future the present's fairness obligation extends. Are there obligations only to the next generation? The next two generations? Indefinitely?

3. *The typology of effects question.* Does every possible impact that a generation has on the future contain moral implications? Which actions must be examined for their intertemporal effects, to be fair to the future?

4. *The ignorance question.* What must one know about future generations in order to be fair to them? Must one be able to predict what they will want and need in order to plot a sustainable course of resource use?

Having separated these four questions, it is now possible to examine answers to them that are apparently implied by the embrace of weak sustainability and the grand simplification on which it rests. Indeed, weak sustainability offers a clear answer to the *tradeoffs question*: its advocates treat this question as the central and controlling question. By assuming that one can compare the opportunities for achieving per capita welfare across generations in a common currency, weak sustainability implies that current generations should only sacrifice current consumption for the future if, and to the extent that, people of the future would otherwise be worse off than the present. Should one be happy with this answer to question 1?

I think not, because this approach to the question assumes that all social values will be adequately recorded by observing opportunities to consume. To reduce the entire question of how to live sustainably to a question of income and consumption levels raises the central question of the nature of environmental value. The weak sustainability answer stands simply as an assertion that one need not consider noneconomic factors in computing sustainability. For Leopold and adaptive managers, there are values not easily reduced to dollars, such as resiliency of ecological systems or the importance of place-based values. So there are good reasons to question the reductionism involved in this solution to the tradeoffs question (McShane et al., 2011).

Weak sustainability theorists also assume an answer to *the distance question* because they perform their calculations after correcting for inflation and other factors affecting the price of economic assets over time. Advocacy of weak sustainability as a guide to decisions thus apparently requires the setting of a "discount rate." Since humans usually favor the present in decisions—seeking to obtain benefits sooner and to delay negative consequences until later—present actors will "discount" delayed effects of their current decisions. To take this time preference into account and to equitably compare current with future consumption possibilities, economists "discount" effects occurring in subsequent years by some per annum percentage. Any significant discount rate will reduce the current value of long-term impacts in comparison to short-term impacts, and may virtually obliterate any impacts beyond the present generation. Such an answer to the distance question seems to ignore impacts of long-term effects, including effects of radioactive waste and severe climate change. In other words, this approach assumes a "presentist" stance by which the sustainability relation can be fully understood as a one-generational transaction. As global environmental problems threaten, this short time frame for calculating weak sustainability seems inappropriate to understand true sustainability.

Weak sustainability advocates also provide, at least implicitly, an answer to the *typology of effects* question. If thinking independently of the economic viewpoint by which all impacts can be understood in terms of effects on individual welfare, it would seem intuitively obvious that some human activities have no real relevance to sustainability, while others are very important to it. For example, if I own a woodlot and cut trees at the rate of growth for the trees, and then replant healthy replacement trees, such activities are—since we rejected absurdly strong sustainability—apparently not at issue in sustainability judgments. If, on the other hand, I clear-cut the woodlot even on steep slopes, fail to plant replacement trees, and allow severe degradation of the topsoil through erosion, this action raises questions regarding the sustainability of my activities. For the weak sustainability advocate, such questions are not relevant unless it can be shown that the clear-cutting had a measurable negative impact on economic capital. If, by clear-cutting, I make a greater profit than by selective cutting, then clear-cutting and its negative impacts would be more sustainable than the alternatives (Clark, 1973). Once again, we see that the grand simplification obscures important questions that seem highly pertinent to considerations of sustainability.

At the very heart of the argument for weak sustainability is the answer that its advocates provide to the *ignorance* question. For example, the economist Solow provides an exemplar of a weak sustainability theory, having stated that sustainability "is an obligation to conduct ourselves so that we leave to the future the option or capacity to be as well off as we are." He also makes it clear that this option can, in turn, be protected by ensuring that the general stock of economic capital of the society is maintained across generations: if future people are as rich as we are, then they will have the resources necessary to create technological innovations that will support adequate consumption and welfare levels. Then, he offers his argument that weak sustainability is the only version of the concept that makes sense, and he offers what can be called an "ignorance argument": "If we try to look far ahead, as presumably we ought to if we are trying to obey the injunction to sustainability, we realize that the tastes, the preferences, of future generations are something that we don't know about" (Solow, 1993, p. 181).

This is a remarkable argument. First, it is hard to ignore the extreme generality of the premise that suggests present people do not know *anything* about what people in the future will want. Does Solow actually mean to claim that we do not know, for example, that people in the future will prefer clean, fresh water to polluted slime ponds? Even if we pass over this apparent gaffe, we can ask: why does Solow think information about preferences of future people should seem important to advocates of sustainability? The answer is that he counts value in terms of individual human preferences, which, once aggregated over individuals and corrected for consumer surplus (the difference between what consumers actually pay and what they would be willing to pay), represent the welfare of the society. Since he has assumed that intergenerational fairness is measured as the capacity of subsequent generations to achieve undiminished welfare comparable to that of prior generations, he criticizes a straw man that would make identification of a path toward sustainability depend on, first, predicting what people of the future will want, and second, trying to provide specific items that will fulfill those wants. Solow thus reaches the conclusion that, since present people cannot know what people of the future will want, and since he has assumed that what is owed them is the ability to fulfill their preferences, the only way to measure sustainability is by identifying it with nondeclining welfare levels. If we question or reject the economic conception of environmental value, Solow's argument against stronger forms of sustainability—ones that specify stuff that must be saved—collapses.

So, now it is clear that the arguments usually put forward to support weak against strong sustainability are question-begging. They are only plausible, based on the assumptions that Solow draws from his economic paradigm that reduces environmental value to a matter of human welfare. To see this, consider the following reversal of Solow's argument based on the same ignorance premise:

1. We are ignorant of the preferences of future people.
2. If the choice to live sustainably is to be based on reasons, those reasons must be available to choosers at the time when they make their choices.
3. Since the preferences of future humans cannot be known in the present, if present current sustainability decisions are to be rational, they must *not* be based on abilities of future people to fulfill their preferences or the comparison of aggregated preferences.

Therefore, the economic, preference-based model used by Solow to discuss current decisions affecting sustainability is inappropriate to the task of charting a course toward sustainability.

But this argument, besides undermining Solow's confident use of mainstream economic growth theory to discuss (and disparage) strong sustainability efforts, raises the obvious question: to what *should* present actors look for guidance in choosing a path to sustainability? The answer proposed here is that guidance should be based on the durable and aspirational values of the community; that is, what the present should save for the future must depend on what is valued today and expected to be valued in the future.

These values must be articulated in a public discourse, a pluralistic dialogue in which participants appeal to the well-established values of the culture and community in question. Achieving sustainability could never—as Solow correctly argues—be a matter of predicting what people of the future will want and then ensuring that they can fulfill those specific wants. Sustainability for a community located in a certain place will be achieved when the values expressing the adaptations and culture are expressed, negotiated, and protected. Sustainability is about projecting past and present values into the brave new world of the future. Or, as was noted following O'Neill, Holland, and Light, sustainability "is about negotiating the transition from past to future in such a way as to secure the transfer of . . . significance" (O'Neill et al., 2008, p. 155–58, quoting Holland and Rawles, 1994, p. 37).

It helps to understand the logic of this situation to introduce an idea from the English philosopher J. L. Austin, who argued that many discussions of ethics get off track because participants commit what he called "the descriptivist fallacy." Austin noted that the grammar of most moral judgments shares its syntax with descriptive statements, as in "Theft is immoral," which has the same form as "Coal is black." We can see that Solow's approach to sustainability, for example, treats the question of what to do as a question that can be resolved by determining, empirically, the effect of actions on human preference fulfillment. But perhaps sustainability is not based on facts about human preferences at all. As Austin says, overemphasis on the descriptive aspects of a situation may cause a failure to see the crucial function of voluntary *commitments* that can be articulated as guides to the future (Austin, 1962, pp. 5–7).

Austin pointed out that there is also an important "performative" function in ethical discourse: citing such utterances as "I do" or "I hereby christen this ship the HMS Morality," Austin argued that accepting commitments, much as in marriage vows, can be an essential function of moral discourse and moral activity. We reject weak sustainability, then, not just because Solow's positive arguments for weak sustainability fail, but also because I think Solow has mistaken the logic of the situation and committed the descriptivist fallacy. Sustainability is a goal to which communities *may commit themselves*; it is not a *description of a state* in which people are assured that they can fulfill their preferences.

If one wants guidance in choosing a path to sustainability, one should consult the trajectory of the time-tested values of the community. One should expect that the community's commitment to sustainability will have the logic of a "community performative," not of an empirical dispute about how to fulfill the most preferences at different points of time. A public discourse, including a discussion of what opportunities the public feels are too important to be lost in the rush toward change, is a discourse about past and present values. In this discourse, features of the landscape and the place provide important support for opportunities that are essential for the community to sustain itself as a cultural unit adapted to a particular physical place.

Weak sustainability is too weak because it limits guidance to those features of the environment that are essential to maintain economic growth. A truly democratic discussion of sustainability will be open to all affected parties, and these stakeholders can be expected to express their concerns in many different languages, including economic, but also in ethical appeals. While such a discourse may be chaotic, it is nevertheless the appropriate starting point when a society faces wicked problems such as identifying a path toward sustainability. It is precisely such a chaotic discussion that can ensure that the full range of values are expressed and considered. Weak sustainability advocates, by defining environmental values in narrowly economic terms, jump too quickly from the messes we face into a single, economic model for measuring value.

Sustainability is about a range of values, values that emerge as one views the system at different scales of time and space. Rather than cancelling out all these scale differences by correcting every temporal value to present dollars, strong sustainability advocates should recognize that, while microeconomic analysis has important functions in understanding individual behavior of consumers, that behavior can also have impacts at larger and normally slower-changing scales of the system. Learning to "think like a mountain" is learning to respond to particular values within a time frame appropriate to the pace of the dynamics involved in supporting those values. Weak sustainability is too weak, to summarize, because it attempts to account for values that emerge at multiple scales on a single, present-oriented scale. The wisdom embodied in this conclusion is announced by Adapt as heuristic 3:

Heuristic 3
The normative sustainability heuristic. Engage the community in a search for enduring values constitutive of its identity. While economic concerns and criteria are important to any community, truly normative sustainability requires a public discourse and the emergence of a "public" that can identify key values that must be sustained if the public interest is to be sustained.

4.3. Scale, Boundaries, and Hierarchical Systems

The concepts and terms that drift into public discourse from technical discourses are often wedded to particular paradigms and assumptions of particular disciplines. To use any of this disciplinary terminology without explanation is to ignore the disciplinary turf wars that are guaranteed to confuse rather than clarify sustainability discussions.

There is a long tradition in conservation thought, clearly emergent in the ideas of John Muir, of thinking about nature and our impacts on it in terms of systems. This insight remains at the center of progressive thinking about environments even today, and it remains the most important intellectual and conceptual counterbalance against reductionistic thinking. For example, while economists recognize that nature is a system productive of many goods and services, they nevertheless model it as simply the resource-exploitation sector of an economy, and they pay little attention to outcomes and variables that are not known to impact economic productivity and the single key variable, human welfare, measured as the fulfillment of human preferences. The attractiveness of this reductionistic model can be seen in the readiness with which Aldo Leopold originally focused attention on deer for humans to hunt and the economic effects of tourism. But careful observers of nature like Muir and, later, Leopold knew that natural systems are more complex than that.

Fortunately, a great deal has been learned about systems and systems' functioning, since the early days of conservation, as ecologists have developed a more dynamic way of thinking about natural and human-made systems. Announcing conservation's holistic roots, Muir wrote, "When we try to pick out anything by itself, we find it hitched to everything else in the Universe" (Muir, 1911). It turns out that Muir was correct, but his statement only reflects one side of the more complex story: not all things are related to the same extent, and impacts in one area of a system may be buffered from changes in other areas by the subsystems that subdivide the larger system.

This buffering effect provides the other side of Muir's story of interconnections. A commonplace example illustrates this point. Consider a lake in which fish populations are newly exploited because of increases in the nearby human population. Systems thinking allows us to view this situation as an interaction across systems—one ecological and one human—but systems thinking also encourages us to understand the combination of natural and human dynamics as forming a larger system encapsulating both as subsystems of a larger, mixed human and natural system.

Understanding of this system will not be served by ignoring the boundaries between the subsystems, because these remain important within the larger system. Suppose, first, that the human population surrounding the lake, initially small, continues to grow. As the small number of humans initially exploits fish populations, there will be little or no permanent effect on the fish populations in the lake. Biologists recognize, though, that in each generation the fish will produce more offspring than lake resources can support, creating competition for survival within the fish population. Up to some point, the effect of human extraction will be negligible: each fish taken by humans will reduce this competition, allowing another juvenile to reach maturity. The human take, at first, has little or no effect on the fish population, which functions as a semiautonomous subsystem. If, however, the human population increases rapidly, eventually they will take more fish than can be replaced by the natural functioning of the system, and larger and larger takings will eventually change the ecology and functioning of the lake system, with unexploited species crowding into the niches of exploited fish. Eventually the lake might drastically change both species composition and the natural functioning of the system.

Does this show Muir was mistaken? No, he was right to emphasize complex relationships, though the other half of the story is that connections among events are often buffered up to some point. Ecologists thus recognize thresholds, activities that—to use the muffin-tin analogy again—stay within a given range of behavior; but they also recognize that these buffering behaviors have limits beyond which normally independent elements of the system spill over and cause negative impacts on other parts and scales of the system.

So, true systems thinking goes beyond the truism that all things are related to each other to the much more complex and confusing idea that all things are (ultimately) related, but that those relations are somewhat compartmentalized and variable in response to thresholds. Further, all these relations are arrayed within a complex of multiscaled, nested systems. Smaller subsystems in this array are often able to function within the conditions (or constraints) imposed by surrounding systems, and they vary significantly at a fairly rapid pace (say, in response to the fishers' fluctuating success rates), even as the impacts on the larger levels of the system are "damped out" by feedbacks within the system.

The two-sided story also explains why integrating ecology and economics proved so difficult. Economic models are limited to a closed subsystem at equilibrium, and in this sense economists operate in a "closed system" of production and consumption, as represented in figure 4.2. As ecological economists such as Herman Daly explain, mainstream economic models of production and consumption do not take the scale of the economy into account (Daly and Cobb, 1989). This model if combined with a goal of economic growth and an assumption that decision makers always act selfishly, guarantees a tragedy of the commons as many individual decisions spill over and damage larger-scaled systems.

Aldo Leopold, based on his experience with wolves, transcended John Muir's belief in simplistically holistic systems and pointed the way toward a more complex and appropriate understanding of systemic interactions. Leopold recognized that cumulative impacts can interrupt or distort normally slower-paced changes in both ecological systems and the abiotic systems on which they depend. One conclusion that Leopold explicitly drew from his analysis was that the land itself is a system, and any success in understanding the land will recognize its systemic nature—the interconnection of

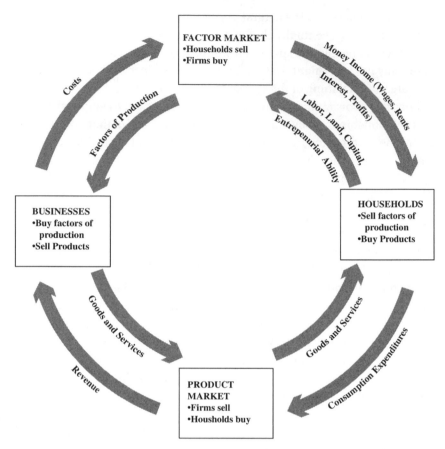

Figure 4.2. The circular flow model of an economy: Diagrammed in multiple ways, this circular model is taught as standard in basic economics classes. This model reduces nature and natural products to "factors of production" and directs attention away from the scale of inputs (resources) coming into the system and outputs (waste) from the system.

its parts. Leopold's productive simile, "learning to think like a mountain," as well as the scientific and evaluative reasoning involved in grasping multiscalar impacts, can thus be understood as instructions for building a more inclusive managerial model, one that includes drivers at multiple levels by tracking multiple dynamics on those levels.

Leopold's use of metaphors and similes pointed toward a systems analysis and, from about 1926, he began to understand managerial problems using models from systems ecology, many of them borrowed from his friend the eminent British animal ecologist Charles Elton. While this period in the history of ecology was tremendously productive and Leopold contributed to it, he was developing his metaphors without quantitative or even conceptual models of what he was talking about in metaphorical terms. Since Leopold's death, ecologists and general systems theorists have developed a much more precise characterization of scalar layering, and thus they have provided what Leopold lacked—a conceptual framework for understanding multiscalar systems and for analyzing them. As I have argued elsewhere, Leopold's simile of thinking like a mountain anticipated the ideas today referred to as hierarchy theory (HT) in ecology, ideas that were developed in general systems theory (GST) and in systems ecology (Norton, 1990).

Today, ecologists, ecological economists, and advocates of more progressive environmental policies can appeal to this sophisticated theoretical framework for the study of systems. Systems theory begins with the assumption that the systems in which humans live and study are complex and dynamic and that much of their behavior is nonlinear.

Here one can see the embodiment of the bright line between weak and truly strong sustainability, between columns 2 and 3 in table 3.1. In surveying the conceptual geography of current sustainability discourse, I introduced and justified a distinction between theories of sustainability that count "welfare" and those that also count "stuff." This is key, because both weak sustainability and strong economic sustainability calculate sustainability within a value system that accounts for nature only by assessing the impact of natural production on goods and services derived within an economic system. The stronger, ecologically oriented positions that appear to the right of the bright line not only count welfare, but they also assess changes in the physical systems that support economic activity and its function in supporting human welfare.

While this is an essential distinction, and one that remains at the center of the theory developed here, the term "stuff" may be a little misleading. One might propound a truly strong form of sustainability by, as noted in the above example, insisting that people of the future should have a certain allotment of fresh water. Failure to protect that amount of water would be, according to this hypothetical, truly strong sustainability theorist, a failure of sustainability; it specifies a measurable feature of the physical system that must be in a specifiable state in order for earlier generations to be able to claim to have lived sustainably. So, as a theory of strong sustainability, this theory would fall on the right side of the bright center line. Because it requires (perhaps in addition to requiring maintenance of wealth) that a given generation protect "n liters of clean, fresh water per capita per year," it would qualify as strong sustainability.

While this hypothetical version of sustainability qualifies as strong sustainability, and is hence justifiably placed to the right of the center line, I certainly do not mean to suggest that "stuff" is limited to objects and materials that people in the future might need. The shift to also counting stuff, from considering the single variable of human welfare as definitive of sustainability to including stuff, has almost always involved an explicit emphasis more on systems and processes than on the particular "stuff" (the stocks) that could be specified.

Nearly all truly strong sustainability theorists thus combine an emphasis on the importance of paying careful attention to physical processes and usually explicitly endorse a form of systems theory as essential to successful characterization of those key variables that must be protected if a society is to live sustainably. Stuff, in other words, includes not just collections of objects in an inventory but, more important, a model of how those objects are generated—the flows—and how the systems that generate them function through time. The physical system, then, referred to in most theories of truly strong sustainability is a system of processes—what ecologists and other systems thinkers call a complex, dynamic system. Our environment is a system; strong sustainability theorists set out to describe what it is about these dynamic systems that is essential to sustainable living. In the remainder of this section, I survey the key concepts that lead to concern to protect key systemic features that are essential to achieving truly strong sustainability: first, resilience; second, self-organization; and third, hierarchical organization.

4.3.1. Resilience

Systems that respond to disturbance and, over time, regain something like the prior path of development that the system was exhibiting before the disturbance, are described as "resilient." Measurements of system resilience exhibit both linear change as a system fluctuates within a given basin of attraction and nonlinear behavior as disturbance thresholds are crossed. When disturbances grow more extreme, they can cause the system to break down and reequilibrate in a new basin. Strong sustainability theorists generally embrace resilience as an important feature because the complex, dynamic systems embodied in their models can equilibrate at multiple regimes of functioning. Management for maximization of a single variable—the optimization strategy—whether for deer or for wheat, tends over time to make systems more "brittle" and vulnerable to flips into another "basin of attraction" (Gunderson and Holling, 2001; Walker and Salt, 2006). This complexity explains why system behavior under management is somewhat predictable and yet can also be startlingly surprising.

Resilience was originally introduced as a descriptive concept in ecology as C. S. Holling (1973) used the concept to describe the ability of systems to rebound after disturbances, which he represented with an equation predicting a particular time of recovery or a particular tendency to recover. Current use of the concept to understand mixed human and natural systems posits multiple, somewhat stable points representing different basins of attraction within a complex, dynamic system, and goes on to describe some of these stable states as more favorable for human use. Ecologists have thus freighted the originally descriptive term, "resilience," with normative content as well. As noted above, one might criticize advocates of strong ecological sustainability for failing to make these evaluations more explicit; it seems that any attempt to articulate a truly strong sustainability will necessarily assign value to processes, especially including those that embody the system's capability for resilience. So perhaps the difference between strong ecological sustainability and normative sustainability merely represents the extent to which the expositors and advocates of particular definitions of strong sustainability explicitly recognize that their ecologically derived concepts, when brought to bear on policy, embody value commitments.

4.3.2. Self-Organization

Complex dynamic systems are maintained and also change and develop through their ability to self-organize. Speaking technically, self-organization refers to the characteristic of subsystems that maintain themselves by capturing energy from the larger system and by dissipating some of that energy back into their environment to sustain themselves through time (Prigogene and Stengers, 1984). Since this maintenance takes place against the background of a changing environment, systems with diversity of responses to environmental conditions can respond by creating new structures, diversifying, and "learning" new responses. These contrasting regimens of maintenance and unpredictable change result in nonlinear behaviors that can only be described in nonlinear equations.

While self-organizing systems are all around us, amazingly, they have only been the focus of scientific and conceptual attention for less than half a century. The reluctance of scientists to engage in the messy mathematics of nonlinear systems delayed for decades the full acceptance and careful analysis of self-organizing systems. Ecological systems and the varied degrees of resilience associated with them account for the complexity, the varied rates of change, and also the mix humans encounter between stable and highly variable aspects of those systems. Understanding how self-organization is crucial for understanding complex systems puts resilience in a central place in discussing sustainability. Resilience is a measure of how far a system can be stretched before breaking. In other words, the resilience of a system represents the stable aspect of systems, while the limits of resilience point to the bifurcation points at which a system undergoes change that can reconstitute the system in a new mode of functioning.

While such system changes are in one sense "neutral"—a flip into a new system might in principle have a neutral or positive impact on human values—ecologically minded managers generally worry that these changes will, more often than not, be taken as negative. Some of this worry may be due to measurable changes that occur in the productivity of a system, but the expectation of negative impacts of a flip are often anthropogenic: as human societies and cultures coevolve with functioning ecosystems, they learn to exploit those products and features that are valued by humans and readily available in the environment, especially those that have economic value. Those cultural adaptations to opportunities supported by the ecophysical system become essential aspects of the economic system as infrastructure and industries develop. If those products and features that support an economy highly integrated into ecosystems become scarce, humans will suffer, at least during a phase in which the economy is reorganizing and before new economic ventures are established. Structural changes in a system might, that is, have negative impacts more because of the human systems developed to use and exploit the system than from a biologically measurable decline in biomass productivity. For example, in the collapse of the groundfish fishery in the Northern Atlantic, overfishing of economically valuable species including cod, halibut, and flounder severely depressed those populations and has led to an invasion of skates and rays. The productivity of the system in pounds of fish flesh has changed little; but the short shelf life of rays, and the lack of nearby markets for these species, means the flip into a system dominated by skates and rays is an economic disaster quite independent of the measurable biomass productivity of the fishery.

4.3.3. Hierarchical Organization

Central to this new, more complex understanding of natural systems and human impacts on them is the idea of hierarchical organization of systems. This idea was anticipated by Leopold when he realized that managing species populations on the "mountain" (ecosystems) requires a systems approach. He recognized that subsystems (e.g., deer-wolves-hunters), when manipulated for human purposes, can, if altered drastically (say, by removing all the wolves), set in motion significant changes in the larger system.

The hierarchical organization of complex dynamic systems supports— and is supported by—dynamics that unfold over multiple scales of time, with fast-changing variables providing the drive, while slower-changing variables manifest at the level of larger systems and provide the constraints under which the smaller systems operate. Resilience of the system (slower fluctuations of large, environmental variables) provides a stable background against which changes in smaller systems create new features through self-organizing activities. One can think of the processes and semistable systems as providing "opportunities" for creative responses. Opportunities, so understood, are the choices open to organisms and populations, given the constraints imposed by the larger environmental systems. In this way, hierarchical organization makes self-organization and resilience understandable and provides a useful systems-based model for thinking about human impacts on multiple system scales.

As one species among others, it can be said that human individuals and human societies face choices and that those choices are mediated, first and most directly, by economic processes that offer both opportunities (open choices, given the current state of markets and infrastructure) and constraints (prices of goods and services compared to income) on those choices. Suppose that one follows ecological economists and integrates economic systems into multiscaled ecological systems (with individual choices within a market forming a process that is both constrained by, but also capable of destabilizing, those larger constraining systems within which an economy develops). One then has an abstract and possibly formal means of representing how human, individual actions, when modified and often amplified by economic markets, can set in motion changes in the functioning of ecological systems whereby smaller, rapidly changing variables have spillover effects on the usually slower-changing variables in the larger systems.

The interplay between fast-changing, driving variables of human economic choices and actions, and the slow-changing variables tracking larger, environmental systems provides an excellent model for articulating and evaluating impacts of human activities on environments. This model, besides integrating economics into larger ecological models, explains both why exploiters of natural systems are so strongly inclined to see them as stable producers of raw materials for economic activities (reflective of the resilience stored in the larger, slower-changing systems) and also why "flips" into different functional regimes can seem to be economic disasters as the system no longer produces the raw materials that support existing industries and the markets they serve.

We can now see why Leopold's introduction of three time scales in "Marshland Elegy" and the way he used this construct in his multiscalar critique of his wolf policy in "Thinking like a Mountain" is so important (Leopold, 1949). While Leopold never used the term "hierarchy," he employs a hierarchical, multiscalar model to show how humans who pay attention only to the outputs from short-term, driving variables in a system for economic gain can actually be undermining the systems' ability to continue to provide those outputs by undermining dynamics that function at a larger level.

It is now possible to introduce heuristic 4, which encourages practitioners of ecological management and participants in collaborative adaptive management to pay careful attention to the multiple scales at which ecological systems change.

Heuristic 4

The scale and boundaries heuristic. Pay attention to scale and self-organization, both in physical dynamics and in political institution building. Getting the scale right is an essential aspect of problem formulation. Whoever is affected by decisions within boundaries of the system bounded for management should be heard.

Taking scalar organization into account is essential but, historically, the tools have been lacking. Hierarchy theory (HT), developed by theoretical ecologists since the early 1980s, thus provides a tool for characterizing and clarifying scalar relationships in systems under management (Allen and Starr, 1982; O'Neill et al., 1986). Formally speaking, hierarchy characterizes a set of nested systems that maintains its functioning over time by a complex interplay across levels, as systems change and reequilibrate against a relatively stable background. Recognition of these slow and subtle cross-scale impacts allows the development of multiscalar models for recognizing and relating subsystems and system scales to each other and allows the use of this model to analyze emergent environmental problems.

By assimilating Leopold's metaphors and representations of his reasoning to a form of protohierarchy theory, I can incorporate the insights of philosophers and scientists who examine the world from the viewpoint of systems theory, as illustrated in figure 4.3. Today, thinking like Leopold but with more formal conceptual apparatus, one can sharpen his metaphorical thinking by incorporating HT. Most adaptive managers, including Holling, Gunderson, and Walker, have explicitly embraced hierarchical thinking as essential to their understanding and managerial initiatives.

HT defines a family of models that include the following attributes:

1. Systems are embedded within one another such that smaller subsystems are expected to change more rapidly than do the larger systems that they compose.
2. All observation of a system must be from within the system. No appeals are made to an extra-systematic perspective.

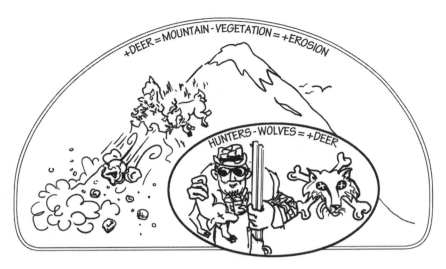

Figure 4.3. Leopold and the mountain: In the early years of wolf extermination, Aldo Leopold bounded the system he managed as deer-wolves-hunters. He later learned that this system is open and interactive, with the larger system as growing deer populations affect the larger and slower scales ("the mountain"), damaging vegetative cover and topsoil development.

These "axioms" define a set of models that are designed, one hopes, to represent space–time relationships in complex dynamic systems. Hierarchical models, thus, provide a delimited, but nevertheless infinite, fund of systems and models that can be used to represent natural systems and also coupled human and natural systems. The trick is to develop models that arrange various dynamics driving a system in ways that increase our understanding and allow us to parse problems onto systems, and systems onto problems, in helpful ways.

These models are tools; the hard part is to characterize an environmental problem so that it can be represented with one or more hierarchical models that illuminate the complex situation in which environmental problems are encountered, complete with ecological, cultural, and political aspects. What is adventuresome in this approach is that, given the type of integrated analysis pursued here, human problems—understood as failures to protect an important human value—can and often do function as the context for choices of models.

HT was developed mostly by ecological scientists and, while most of them agreed with my implication that hierarchical models are constructed as tools, not as representative of the actual structure of the system involved (Allen and Starr, 1982), it seems they have not paid as much attention to the ways in which political and policy issues, and also social values, affect perception and what is of interest in particular situations. Thus, while many hierarchy theorists speak as if they choose the scales and boundaries of systems they analyze based on standards of good descriptive science, it is no less true that models can be—and have in some cases been—constructed with a clear eye on important social values. Since HT is merely a tool for organizing space-time relationships, one is free to design or choose a particular hierarchical model that seems appropriate for modeling a given environmental problem.

So understood, this tool cannot solve the problem of choosing an appropriate multiscaled model; it merely provides a vocabulary and a basic structure for building useful models that will illustrate important relationships and make the scale choices explicit. The appropriateness of the model is at least partly determined by the goals and values that motivate action. Hierarchy can thus be used as a tool that can contribute both to develop useful scenarios that track development paths into the future and to explicitly represent and track important social values.

Structurally, because HT tells us to expect larger systems to change more slowly than their subsystems do, from the middle, focal level, one can look upward to larger systems to encounter and understand constraints acting on the focal level. If one looks downward toward the subsystems of the focal system level, one can identify and characterize mechanisms that interact to drive behavior at that focal level. Understanding the three levels together, one can understand the patterned dynamism of complex systems.

Applying the tool to Leopold's analysis of what went wrong with his wolf and deer management, it can be seen that he at first developed a model based on the metaphor that nature is a productive system that can be managed to maximize outputs of desired products (such as deer to hunt). Building on this metaphor, Leopold examined the workings of the productive system and, given the productivity metaphor he assumed, he concluded that the system could be more efficient in producing deer if competitors for deer were eliminated. Leopold had, in other words, fallen into the use of an optimization strategy that would make Optim proud. Because he concentrated on only one product, deer, and because he did not look for other social values that could have caused him to think twice about maximizing the deer-for-hunting variable, Leopold was unpleasantly surprised to learn that hunters could not control the boom-and-bust cycle of predatorless grazers.

In the terms of HT, Leopold created a truncated system model that reflected human actions and impacts on the size of the deer herd (which normally fluctuates in response to local conditions, but usually "balances out"). Yet he did not recognize that his actions to increase deer would also affect larger-scale and normally slower-changing variables, such as vegetative cover and depth and stability of topsoil. If he had been thinking hierarchically, Leopold would have recognized that the background conditions—vegetation and topsoil—would eventually be affected. These larger geological and ecological systems actually placed constraints on the size of the deer herd, constraints Leopold missed because he focused on too few levels of a complex system and too short time frames. As a result, he set in motion changes in the usually slower-changing variables of vegetative cover and composition of soils, changes that were to reverberate through the system and to enforce the constraints implicit in the larger systems. Enforcement was in the form of a huge die-off of deer in a hard winter.

Leopold's simplified model led him into a serious blunder: he recognized that his actions in service of economic values had damaged other important values associated with maintaining vegetative cover and protecting mountainsides from eroding into streams. He later adopted a more complex and appropriate model, the multiscalar model of deer, wolves, hunters, and vegetation all evolving together on a mountain. Once he saw that the simpler model missed important impacts on the system, he realized that important values will be ignored if his productivity model dominated other models.

While Leopold never stated the axioms of HT, I believe he can be credited with the concepts and the ideas that today form the theoretical core of HT. In fact, the ideas of HT had percolated in biological discourse for several decades before these ideas were given a more formal framework (Allen and Starr, 1982), and it is not surprising that Leopold appropriated and contributed to this line of thinking. The connection of HT to adaptive management theory has, since Leopold's day, been cemented by the appropriation of hierarchical modeling into adaptive management theory (Holling, 1992; Walker and Salt, 2006).

Building on the insights of Leopold's discovery of the importance of scale and of the elaboration of these insights in adaptive management, HT is here introduced as a flexible tool for constructing models of local situations and problems. As such, its core is an extremely simple structure of nested systems, though its application in particular situations will depend on a sincere attempt to build one or more hierarchical models that clarify management choices. So, HT and hierarchical modeling simply provide a set of concepts that help construct model systems.

What I have added here—and it may be controversial—is that hierarchy can be an especially useful tool in environmental management, not merely in ecological descriptions. I will use hierarchical concepts to develop models that can be designed with specific social goals in mind, models appropriate in scale and dynamics given the social values being protected in particular situations. Most ecologists would be most uncomfortable with this mixing of science, goals, and values—but not all of them, as will be learned in chapter 8 (Pickett and Cadenasso, 2002).

4.4. A Schematic Definition of "Normative Sustainability"

Any concept or definition of sustainability must be normative in the sense that it specifies, for a given community, a sustainable path to the future, a process by which the enduring and aspirational values important to the community are effectively expressed and protected; the values to be protected in that community will be associated with physical features of the environmental systems in which the community is embedded. A normative definition differs from a descriptive one, such as Solow's definition, in giving evolving community values a central role in specifying sustainability for that community. But, since different communities will emphasize different features of special value, I seek a *schematic definition*. Accordingly, our schematic definition specifies the "form" for normative definitions, leaving important variables—identification of particular features of value—as a key task of each sustainability-seeking community.

I will discuss these two steps in two separates sections: Section 4.4.1 provides a general schema, a definitional form that includes variables that can be specified by deliberation and choice as communities learn a path toward sustainability appropriate to their values. These values will be associated with certain features of the landscape that support and protect opportunities to enjoy the values characteristic of their place and their community. In section 4.4.2, I use HT to show how particular opportunities can be associated with physical features of the landscape.

4.4.1. A General Schema for Sustainability Definitions

While one can learn a lot by speculating and arguing about the true meaning of sustainability, it will not be possible to provide a universal definition. As noted, one of the problems is that sustainability—or lack thereof—can be attributed at many different scales: individuals, practices, regions, and countries can all be said or denied to be "sustainable." I doubt it would be fruitful to argue whether one or the other is the right scale at which to look for sustainability. While I hope the totality of the discussion in this book disarms any sense of arbitrariness, I am stipulating that our attempt to characterize sustainability will begin from the viewpoint of a local community. Of course, a rich form of sustainability must include a sense of the space around a place. So, a full-blown, community-based concept of strong sustainability would also include concern about larger systems, up to and including the earth's very atmosphere. Such a conception of sustainability is directed toward all scales of natural-cultural systems, though from a local viewpoint.

My sense of community is quite inclusive; in some cases, a community will be a tight-knit unit with clear boundaries and a sense of identity; in other cases, the emergence of a problem or sense of crisis may create a community where there had been none. A community, in this inclusive definition, is understood as a group with a shared identity. More specifically, some communities are connected in important ways with a physical area and the natural characteristics of that area. I refer to such communities as "places"; a place is a community that enjoys a special relationship to a natural system. Since some communities have not perceived the importance of the natural features of their area, one might then speak of communities that have the potential to become a place. They are communities that might develop a shared set of concerns regarding their place.[1]

Having established in section 4.2 that the sustainability relationship requires more than promoting economic business-as-usual, true sustainability can only be achieved if a community commits to protecting "stuff" that supports their way of life and the values they enjoy. The challenge, then, is to characterize a set of goods that are general enough to serve as placeholders in a normative definition of sustainability and able to connect those goods to features, elements, or processes of natural systems in particular places. If we can provide such a definition, then we will have given the idea of sustainability a core meaning—one that is variably instantiated across communities. Such a definition leaves specific "variables" that must be specified as communities evolve toward their own definition of sustainability.

The key to understanding multiple definitions appropriate to specific communities is the notion of an "opportunity." An opportunity is a possible action or policy that could be chosen, given the environment at hand. The key aspect of this set of definitions is the phrase "given the environment at hand," which refers to some feature or aspect of the local environment that is essential to the choice (opportunity) in question. Cutting and milling timber, for example, is an opportunity in a community with healthy stands of trees; if all the trees are cut without replanting in one generation, it can be said that the opportunity that once existed in the place-based community—the opportunity to harvest trees or run a sawmill—has been lost. The important conceptual point is that an opportunity as here defined includes a reference to specific resources or features of the place in question.

With these definitions and understandings in hand, I can now provide a schematic definition of strong, normative sustainability, as follows: *For a given generation to live sustainably means that that generation fulfills its needs and desires so as not to destroy important and valued opportunities for future generations* (adapted from Norton, 2005, p. 363). This definition builds on the idea that individuals and populations living within a system experience their environment as a mixture of opportunities and constraints—and sustainability will protect the opportunities in this mix.

This definition is schematic in that the choice of actual opportunities considered important in diverse communities will vary widely. The definition leaves a role for communities to democratically decide what activities are valued deeply enough to sacrifice to protect them for future generations— what *should* be saved. One should not forget that this approach, which leaves actions, choices, and policies unspecified, as variables, is insufficient as a definition of sustainability for any community. To actually instantiate this definition, it is necessary for a community to engage in a process by which opportunities are associated with the community's deepest and most aspirational values. The task will be to describe a process by which a community, through democratic means and through participation in deliberation and negotiation, might define for itself what sustainability means to their community.

4.4.2. Opportunities (and Constraints) in Hierarchy Theory

I have stressed, in this chapter and throughout this book, that thinking about sustainability is all tied up with scales of time and space. At first, spatial ambiguity and open-endedness of situations seemed to confound understanding of intertemporal morality, contributing to the wickedness of environmental and sustainability problems. As a partial antidote to these problems, I have now introduced HT as a tool to clarify spatiotemporal relations by providing a set of models that are hierarchical in the sense that each model is structured and understood according to the two axioms of HT. Smaller subsystems in a set of nested systems will normally change more rapidly than the larger systems they compose; and all observation, measurement, and evaluation must take place from somewhere within the system.

Thinking hierarchically can evoke interesting and important insights about adaptation, because the Darwinian model of evolutionary change through natural selection can be understood as a hierarchical process. Whenever one thinks about adaptation and selection, one is thinking about an individual struggle in which an organism tries to survive and reproduce in an ecological habitat, with survival depending on being appropriately adapted to the environment in which it lives. Individuals (small systems) live and die within an environment (larger, encompassing systems). Their success in surviving and in reproducing depends on opportunities found in their environment. These opportunities allow choices that support survival in the specific environment in which they live.

Given this abstract representation of adaptation and behavioral choice, each individual in an environment experiences that environment as a mixture of opportunities and the contrary thereof, constraints. These concepts lie at the heart of the hierarchical system introduced here, because they can serve as bridge concepts. By connecting semantically with the value-laden concepts of opportunities as enablers and of constraints as impediments to desired actions, these concepts have an inherent evaluative element; yet they can also be connected to physical features of the landscape that can be described physically. Just as snow provides an opportunity for skiers, and aridity constrains agriculture, one can associate physical aspects and features of environments with a range of choices in the struggle to survive. Viewing organisms as embedded in an environment and facing a mix of opportunities and constraints creates a bridge between the language used to describe the features of an environment and the idea of opportunities that connect to value discourse. This abstract representation can be applied to any scale and to many types of evolution over time.

In the context of adaptive management, however, hierarchical reasoning is applied to choices available to humans as they live and evolve within a partly stable and partly changing environment. Applying the analogy of adaptation to human choice situations, one can characterize these as having a generous mix of opportunities (many resources to develop and use) over constraints (the aspects of the environment that limit choices and lead to failed survival and reproduction). A positive mix of opportunities vis-à-vis constraints for human societies is a necessary condition of what the philosopher-economist Amartya Sen calls "opportunity freedom" (Sen, 1999, 2002). Opportunity freedom refers to the real choices open to people in a social environment. This concept will be discussed in more detail in section 6.3.

In particular, the idea of natural selection can be used to introduce a key term in the schematic definition. The terms "opportunities" and its contrary, "constraints," can be modeled within hierarchical systems. Evolution occurs when the gene pool of the population is affected by the successes and failures of individuals within specific environments. One can understand behaviors and choices of individuals as taking place in an environment with a certain mix of opportunities and constraints. The latter function within the schematic definition as a bridge connecting the descriptions of the physical features of a system and the aspirations and values that members of a community cherish and wish to perpetuate. (See figure 4.4A.)

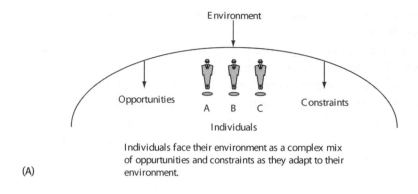

(A)

Individuals face their environment as a complex mix of oppurtunities and constraints as they adapt to their environment.

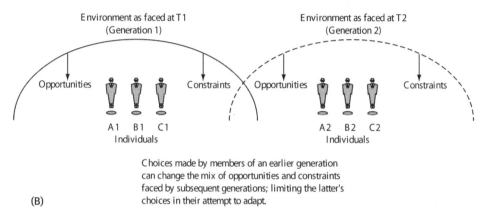

(B)

Choices made by members of an earlier generation can change the mix of opportunities and constraints faced by subsequent generations; limiting the latter's choices in their attempt to adapt.

Figure 4.4. Hierarchy and sustainability: Individuals (whether humans or other organisms) experience their environment as a mix of opportunities and constraints, as in (A). Applying this model, based on HT, to a cross-generational comparison, as in (B), an earlier generation will be considered *unsustainable* if it uses up opportunities and leaves mainly constraints for subsequent generations. Maintaining or improving the mix is *sustainable*.

Hierarchical models, so understood, connect the social values of a community to the physical features of the place in which they pursue and enjoy those values. This model and its bridging features between descriptive language and values associated with opportunities provide a definition of sustainability that is based on the idea of adaptation and, more specifically, on the idea that successful living requires the existence of opportunities that support successful choices. An individual organism is adapted if, in negotiating available opportunities and constraints that are describable in physical terms, it survives and reproduces. But survival of a population in the long run will require also that offspring face a sufficiently rich range of opportunities as they too attempt to survive and reproduce.

Consider now figure 4.4B, which represents the same relationships but compares these relationships across generations. This framework provides background and meaning to the schematic definition, and from it, the following version of the schematic definition emerges: *a set of behaviors is sustainable if and only if its practice in generation* m *will not reduce opportunities in the mix of opportunities and constraints that will be encountered by individuals in subsequent generations,* n, o, p. . . . This little model thus provides the basis for representing normative sustainability as a comparison of opportunities available to a given generation. This model is especially effective in characterizing "unsustainability": an action is unsustainable if it reduces the opportunities or increases the constraints in the mix faced by subsequent generations.

Applying this model to the case of human choosers, an opportunity based in nature is a culturally meaningful aspect of a physical system; viewed from the perspective of the physical system, crucial aspects of that system may be essential for the culturally perceived opportunity to exist. This definition of sustainability encourages communities to identify certain opportunities as essential to the continuation of their culture and also to associate those opportunities with physical dynamics, thereby completing the connection between human goals and features of the physical world. Such a connection is the essential bridge that allows linking human values with natural features and processes so as to define an approach to sustainability that is both strong (it specifies stuff that must be saved) and normative (it is guided by social values of the appropriate community).

4.5. Conclusion of Part 1: A Way Forward

A definitional schema for a normative understanding of strong sustainability has now been introduced. It might seem that this should be the culmination of my argument: that, having a definition, the task is completed. Not at all! My constructivist, community-based definition provides only the beginning point for a deeper understanding of how to successfully pursue sustainability. This schematic definition, while providing a core idea of preserving opportunities, leaves for further specification — or for detailed public processes — identification of the actual opportunities and features that might be filled in by communities facing the future with a sincere desire to chart a sustainable path that pursues those aspirations into the future.

It remains to be learned how appropriate variables are best characterized, identified, and pursued in a discourse appropriate to a given community. While I have been rather hard on the disciplines, remember that my arguments have mainly been directed against disciplinary claims that a given discipline can provide a *comprehensive* treatment of forward-looking environmental problems, and against claims that environmental values can or must be expressed using a single terminology. Having cleared away misleading assumptions, and having adopted a more realistic understanding of environmental values as pluralistic and evolving, we can now begin to build a new approach to integrating values and empirical knowledge into sustainability definitions.

Pluralism, as opposed to monistic, single-value systems of analysis, creates a more open and inclusive dialogue about what to do, and why. All areas of environmental science can contribute to understanding, but the conceptual frameworks offered by the special disciplines will never achieve comprehensiveness over all environmental values. For the foreseeable future, environmental discourse will be fragmentary, with some phenomena interpreted economically, and other phenomena described in more spiritual or philosophical terms. There need be no rush to choose a single, comprehensive framework for evaluation of change, for one can now envision an open, far-ranging public dialogue about management goals and sustainability pursuits, and one can even anticipate that economists, ecologists, and environmental ethicists will all be involved in that public discourse.

For those enamored of the myth of singular answers, this approach will sound chaotic. Yet, from the pluralistic viewpoint here pursued, the openness of the discourse, as well as the diversity of voices, are important inputs into a decision process. Pluralists can set aside the divisive arguments about which system of value provides the one, correct characterization of human values, and they can then focus more on what can be done together and less on disciplinary ideologies about the nature of environmental values.

This way forward will require a revision of the role of the special disciplines in the policy development and deliberation process, creating a new way of understanding public discourse in pursuit of environmental policy goals, including sustainability. It will also require the development of a process of decision making that can advance toward cooperative action while disagreements regarding the nature of environmental value rage on, and even while collaborators endorse and pursue different values, even incommensurable ones. Since I do not seek a single "best" outcome through analysis, the way forward must be to accept, for the foreseeable future, the chaos. Still, this is not a concession to despair; on the contrary, with a clearer understanding of the situation, focus can shift from analysis to process and deliberation, including critical thinking about boundaries, political representation, and the like. The way forward will be to develop a process-based model that is adequate to the needs of adaptive management.

As this chapter—and this, the first part of the book—ends, I now move beyond criticism of failed attempts to evaluate change, and I reject unpromising attempts to force sustainability thinking into monistic and optimizing understandings. In the next part, a procedural approach to understanding environmental problems will be developed.

PART 2

A Process Approach to Sustainability

CHAPTER 5

Introducing and Grounding a Procedural Approach

5.1. Pluralism and Corrigibility

Having proposed proceduralism, and having renounced the search for uniquely optimal solutions, we now shift our attention to the alternative conception of uncertainty and social learning inherent in adaptive management. An adaptive approach to understanding the puzzles of environmental management need not be nonrational or excessively chaotic. Nor need it be dominated by ideology, which leads to unending disagreements. What adaptive managers need is a new approach to evaluating change—one that applies empirical methods to improve processes and thereby promises better decisions.

Assuming that the goal is an environmental policy that is effective *and* democratically legitimate, then the best approach is a pluralistic viewpoint that emphasizes free speech, freedom of expression even for ideas thought heretical, and open discussion and debate of all matters affecting the public. Pragmatists advocate open procedures, with all voices heard, and their pragmatic method operates on the assumption that one can learn from those with different viewpoints. Unlike monists, pluralists encourage and expect to learn from a diversity of opinion; pluralism and free expression contribute to truth-seeking. Supreme Court Justice Oliver Wendell Holmes expressed this view when he wrote, "When men have realized that time has upset many fighting faiths, they may come to believe even more than they believe the very foundations of their own conduct that the ultimate good desired is better reached by free trade in ideas—that the best test of truth is the power of the thought to get itself accepted in the competition of the market, and that truth is the only ground upon which their wishes safely can be carried out" (Holmes, 1997, p. 180).

Even if one rejects Holmes's emphasis on a singular truth to be known only in the distant future, one can accept his call for "free trade" in all manner of ideas that he justifies instrumentally, as the most effective filter for removing falsehoods and serving scientific progress. The role of dissent and the expression of plural viewpoints on the way to improved understanding is a keystone of progressive democracy, and committing ideas to the maelstrom of public discourse, while indirect, eventually calls all assumptions and foundations into question. Pluralism, recognizing the fact of diverse values, viewpoints, and perspectives, is the engine of rational public discourse.

The answer to the charge that pluralism leads to relativism is the same answer that can be given by any scientist: objectivity is sought in a process by which claims can be refuted and disagreements reduced through debate and critical thinking. This means that adaptive management, recognizing its role as seeking to refine both the means and the goals of an ongoing process, is committed to applying the scientific method—to experiment when possible and to observe carefully when not—to all disagreements, including those that involve complex mixtures of factual and evaluative conflicts.

What is needed is a "procedural epistemology"—some guidance regarding how, exactly, to address wicked problems using the scientific method of experience. Speaking more specifically, how can the study of the diverse values of a community rely on the results of our study to chart a course toward sustainable living? Can empirical studies help a community choose appropriate processes for determining sustainability goals? Yes, they can, provided we focus on procedures that can lead to improved communication and cooperative action. Making the discussion of values and goals a matter of public discourse can lead to trust and opportunities.

The task, then, is to provide a process model that is normatively adequate, in the sense that it can, over time, clarify and rank community values in an open-ended adaptive process; it must also be epistemologically efficacious, in the sense that it can guide a self-correcting process of action, evaluation, and realignment of multiple values. These two tasks can be understood, respectively, in terms of two key challenges to the process approach to guiding environmental policy. These challenges can be represented as the search for a normative anchor (which will be proposed in section 5.2) and an efficacious method (the subject of section 5.3).

The first challenge is to provide *an anchor concept* in the form of a process-based ideal that can be used as a basis to compare actual processes with hypothetically ideal processes and to render judgments as to whether those processes are appropriate. The ideal, the public interest, is defined procedurally; it provides a hypothetical anchor that explains how engaging in such a process can aid in the pursuit of evolving and flexible goals.

The second challenge is to provide *a method* that increases the likelihood that communities engaged in such processes will achieve broad-based social learning, which will bring the dynamics of value formation into play in the process by which a community struggles to articulate the public interest.

5.2. John Dewey: Publics and the Public Interest

The public interest provides a beacon that shines light on policy formation and implementation. I will specify guidelines for deliberation and public involvement that will allow communities to evolve and learn their way toward goals and activities that will protect and enhance the public interest. Sustainability viewpoints can be integrated into this idea of the public interest by pursuing a multigenerational public interest. So, the criterion that is used to identify better decisions is whether those decisions protect or threaten the long-term public interest of the community in question.

Of course, this characterization of good decisions is little more than a blank check, as its application requires definitions of "public interest" and "community." In the remainder of this section, I explain the public interest and how it can serve both as a flexible articulation of the goals of management and also as a normative anchor of a process-based approach.

The concept of the public interest had until recently fallen into intellectual disrepute, because of its ambiguity and difficulty of specification, but the concept has been given new life through reinterpretation within the American Pragmatic tradition (Norton, 2005). In a recent, insightful book, Barry Bozeman and Ben Minteer have argued that, with respect to specific problem situations, public discourse and deliberation can in the best cases "construct" a "public interest," in the style of John Dewey (Dewey, 1927; Bozeman and Minteer, 2007). They argue that Dewey's conception of the public interest can be characterized as the outcome of a process that can be understood as broadly applying the scientific method to social problems and controversies. It turns out that, when properly developed, the public interest can be used to fill a bothersome gap in some forms of proceduralism.

As noted above, continuing in the tradition of procedural rationalism, which equates good decisions with those that emerge from an "appropriate process," I treat the search for solutions to environmental problems as the project of specifying appropriate processes in various situations. Better processes yield better decisions, because good processes are more likely to reach agreement on cooperative action. Advocates of the strategy of trusting the outcomes of good processes, however, face a choice.

Pure proceduralism essentially ignores the problem of assessing appropriateness and simply accepts the outcome of an actual process that is in place as yielding the best decision. Whatever decision emerges from existing processes, however flawed, will be considered the rational decision for that group. Pure proceduralism seems, at first, to be the most "democratic" approach, but of course, there are problems with "pure" democracy. Many actual situations are distorted by power relations, and people's interests are often manipulated; unless some constraints are applied to decisions emerging from a process, there is little reason to give normative weight to such decisions.

Constrained proceduralism allows free play of deliberation, accepting its outcomes, though the matters open to decision in the process are restricted. For example, one might adopt the attitude of pure proceduralism—accepting the emergent outcome as validated—within a constrained decision space that is limited, for example, to options that are "constitutional" in their jurisdiction, consistent with the International Declaration of Human Rights or the Rio Convention on Biodiversity. Constrained proceduralism limits the range of decisions open, but (as in pure proceduralism) outcomes of existing processes are taken as democratically justified simply because they were chosen in an existing process that is more or less legitimate.

While, broadly speaking, these approaches seem democratic, most would find these pure or constrained specifications unsatisfying because, even though they allow free choice over a given range of decisions, neither provides a plausible explanation of why we would expect that decision to be better than any other. In this sense, pure procedural theories are "hollow at the core." Once a group has rendered a decision, one can only say that it is good because the group made that choice.

Procedural rationality can be saved from vacuity and circularity, however, by introducing *heuristic proceduralism*, which bestows legitimacy on particular, identifiable processes by progressively specifying procedures likely to be useful in solving problems at hand. These heuristics are hypothesized (and tested) as characterizing productive decision processes in which open deliberation can create, from diverse voices, an agreement that has withstood critical attention and empirical tests as a part of the process. One can put trust in deliberation and define "appropriate process" by specifying characteristics of an idealized public discourse and decision procedure that encourages social learning. A process is appropriate if it tends to encourage agreement and cooperative behavior within a deliberative process. It is useful to refer to the work of the second-generation pragmatist John Dewey to explain this process. Dewey's dream was a modern society in which individuals flourished by developing deliberative skills that allow them, individually and collectively, to apply the method of experiment and observation and to adjust policies accordingly. Dewey's thought is thus relevant to the current discussion of a democratic process, and it will inform the understanding of adaptive management in this and the subsequent chapters.

Dewey described the democratic process as follows: "I would not claim that any existing democracy has ever made complete or adequate use of scientific method in deciding upon its policies. But freedom of inquiry, toleration of diverse views, freedom of communication, the distribution of what is found out to every individual as the ultimate intellectual consumer, are involved in the democratic as in the scientific method" (quoted in Bozeman and Minteer, 2007, p. 108). This approach avoids vacuity not by specifying in advance a "better" outcome using a substantive evaluative tool such as cost-benefit analysis (CBA), but rather by stating an ideal of public discourse and then characterizing a decision as "better" insofar as it was arrived at by processes closer to the ideal. Following Bozeman, I adopt a Dewey-inspired strategy that has been developed through communicative ethics in Europe and that now represents a standard approach in social sciences.

Bozeman and Minteer follow the philosopher Walter Lippmann who defined "the public interest" as what people "would choose if they saw clearly, thought rationally, acted disinterestedly and benevolently" (quoted in Bozeman, 2007, p. 83). Admitting this is an ideal, Bozeman and Minteer nevertheless give substance to the ideal by seeing the public interest as that set of goals and actions that would be arrived at in a democratic process that encourages citizen involvement and social learning: "Dewey's work adds at least two additional critical elements to public interest theory: a method of democratic social inquiry modeled after the ideal workings of the scientific community, and a focus on the key role of deliberation, social learning, and interest transformation in this process" (Bozeman and Minteer, 2007, p. 104).

Bozeman and Minteer indicate the following:

According to many conventional theories of the public interest, especially procedural approaches, there is no claim to identify an invariant and monolithic public interest but, instead, there is a search for a public value or set of public values that serve the public good. The public value quest *starts with the social problem* and then works toward a limited ideal . . . Public interest theory has been faulted for being ambiguous (i.e., having multiple meanings), but in fact the public interest, viewed as public values, *should* have multiple meanings. (Bozeman and Minteer, 2007, p. 100–101)

They recognize that "[t]he public interest, in Dewey's view, is thus not an absolute, universal, or ahistorical good. It is constructed in each policy and problem context as conjoint activity that produces indirect social consequences that the democratic public wishes to direct into collectively identified and validated channels" (p. 108). In this view, the public interest begins as a procedurally specified ideal. The public interest is that which would emerge if that ideal were pursued in specific, locally grounded processes whereby a community, engaged in public discourse, seeks solutions to perceived problems. According to Bozeman and Minteer, "[t]he public interest in the pragmatic view is a contextual and pluralistic good, one constructed in each policy and problem context by a democratic public committed to the cooperative and deliberative process of experimental social inquiry" (Bozeman and Minteer, p. 110). The most salient characteristic of the public interest is that it is not an aggregation of individual interests, but rather represents a communal value (see Norton, 2005, sec. 7.1).

This constructivist project, however, turns out to be more powerful than might be thought. While not conceived as a fixed, describable-in-the-present endpoint, both "the public interest" and "a sustainable future" have in common that they can be defined hypothetically by associating them with the outcomes of appropriate processes. Once it is acknowledged that the goal of adaptive management is to act cooperatively in the democratic pursuit of sustainability through open discourse, application of the scientific method, and cooperative behavior, one can begin to learn which processes are appropriate. Actual processes can thus be compared with an ideal process.

The pragmatic idealism approach of Dewey and of Bozeman and Minteer can be adopted as the conceptual anchor, avoiding relativism and nihilism. The public interest is a moving target, yet experience—systematic experience in the form of experiments and careful observation coupled with actions— can help approximate it, even as it is redefined. Values—and discussions of which ones are important to protect—will be an essential aspect of deliberation and social learning, and this hypothetical public interest can be understood as embodying not only present values but also those values that would emerge from serious deliberations about what is important to save (see section 6.1.).

Armed with the idea of an ideal process, defined as one that is appropriate for a community seeking to act in the public interest, one can find a middle way between the relativism that "anything goes" and the absolutism of one-right-answer positivism. There are many "truths," and the validity of each truth is best determined by judging the manner in which it is pursued, rather than by comparing it to a "real" world beyond experience. The actions of a community are adaptive, provided they address these problems in a way that is appropriate, approximating an ideal process. Such a process, if achieved, will eventually bring the scientific method to bear on disagreements that matter to management choices. While never definitive of a single truth, seeking appropriate practice can, over time and through iterative treatment, uncover and eventually validate better solutions: it is not perfect, but it is self-corrective.

5.3. Heuristic Proceduralism: A General Method

5.3.1. Process Thinking

If it were possible to get past ideological assumptions and their stranglehold on the human imagination, a new approach to environmental evaluation might emerge. To have an illuminating discussion of values and environmental policy, it will be necessary to develop a way of characterizing and evaluating the effects of specific human activities on the processes of mixed natural and human-managed systems and, in turn, on the impact of those effects back on communities.

So what is needed is a way of evaluating *development paths*, which can be thought of as the set of possible pathways into the future. In the discourse of adaptive management, the unit of analysis for evaluation should be a path forward, a process that could unfold if certain choices are made, particular goals set, and specific actions pursued. Pluralists can use a checklist of characteristics by which to compare and rank possible futures. This plural checklist thus uses multiple criteria that will take the form of a set of empirical, indicator measures associated with social values to which the community gives priority. Particular development plans can be associated with various policies and choices the community can take, and these possible futures function as visions that might be pursued under a strategy of heuristic proceduralism. A description of a possible development path into the future can be thought of as proposing experiments with actions that will sustain or further important values. By adopting a process-and-action-oriented, iterative process that can be self-corrective of both scientific beliefs and emerging social values, progress can be made, even on complex problems.

The proposed approach would include, in some form, the following steps or elements:

1. Describe several possible paths into the future, and associate each with a set of policies or actions that would make that path more likely. Here, the emphasis is on creative and speculative thinking in the identification of possible paths and ways to proceed on those paths.
2. Associate those paths with characteristics that can, in turn, be treated as measurable indicators. Indicators are empirical measures of progress in providing or protecting a social value, even though in some cases the measure may be indirect.
3. Generate public discussion of the comparative importance of the various indicators, integrating this discussion with comparisons and evaluation of different scenarios and in the process using and improving evaluative techniques.
4. Guide discussion toward actions that are thought to favorably affect those indicators that score most highly on the importance ranking but that are robust in the sense that they also have good or neutral impacts on other important indicators as well. This guidance suggests it will be useful to develop multicriteria decision tools as part of the decision process.
5. Seek opportunities to find common ground in actions; build coalitions to experiment and explore options. Cooperative action even on small issues can build trust and expand opportunities for further progress.
6. Study outcomes scientifically, employing the method rigorously, and evaluate actual outcomes. Adjust goals and reprioritize values and then return to a discussion of the evolving situation, perhaps with more trust and better abilities to cooperate.

Heuristic proceduralism is normatively anchored in the public interest, understood as a procedurally defined ideal that identifies a flexible goal for the adaptive process. That goal will not be pursued in a straight line; diversity of opinion is expected and welcomed, greasing the wheels of public discourse. Heuristic proceduralism thus specifies, for present purposes, what Simon (1976) refers to as "appropriate processes" in his definition of "procedural rationality." Appropriate processes are ones that follow the heuristics that are relevant to the decisions a community faces.

While the epistemology developed here—that of the scientific method—cannot offer certainty, it does offer corrigibility. If applied persistently, and debated openly, false assumptions and unproductive ideas will be superseded. But, of course, this is not a process with an end; management of dynamic, mixed human-natural communities requires ongoing responses to evolving situations. This follows from the recognition that most environmental problems are wicked problems or have wicked aspects and therefore suffer from the lack of an agreed-on problem formulation. To face that problem directly, I highlight Adapt's fifth heuristic.

Heuristic 5
The process evaluation heuristic. Think of evaluating change as an ongoing process; the goal of evaluation should be a comparison, and an ordinal ranking, of possible development paths into the future, an activity that continues as a community transitions from present to future. A multiscalar approach allows the evolution of development paths over multiple scales of time and space, capturing the idea of pursuing a multigenerational public interest for the community.

Too much pessimism can be associated with wicked problems, because too much has been made of the lack of algorithmic solutions. In fact, rejection of those models opens the door to an alternative, more adaptive, experimental, and process-oriented rationality. The messiness of problems—and the futility of principle-based, ideological principles in addressing them—was already clear to Dewey, decades before Rittel and Webber described wicked problems. Dewey, while emphasizing a problem-based approach, did not see "problems" as well-formed questions awaiting decisive actions; indeed, he was quite explicit in seeing the articulation of a problem as a function of an evolving "public." A problem emerges in public discourse as inchoate and open to many interpretations. He wrote, "Retrogression is as periodic as advance. Industry and inventions in technology, for example, create means which alter the modes of associated behavior and which radically change the quantity, character and place of impact of their indirect consequences" (Dewey, 1927/1982).

5.3.2. The Emergence of a "Public"
Relying on Bozeman's and Minteer's updating of Dewey's approach to social learning and public interest offers a method by which to gradually specify an emergent public interest; while inchoate, the process approach is self-corrigible. Dewey's process begins not with institutions or well-laid plans but rather with the expression of concern for the consequences of actual or possible policies; this originally inchoate concern becomes more articulate and coherent through deliberation, and from this process there emerges a "public" and unified voice that expresses a viewpoint. Perhaps Dewey overemphasized the unified voice, but his understanding of the public interest as emergent from a process of public discourse encourages a focus on an emergent deliberative process. Today, it seems more realistic to amend Dewey to recognize that there will no doubt be multiple publics and their cohesion

will surely be a matter of constant evolution and devolution. In the words of Bozeman and Minteer, "there will be many 'publics,' just as there will be many public interests in various times and places" (Cochran, 1999; Bozeman and Minteer, 2007, p. 110).

Dewey begins with open-ended discourse, not with well-defined solutions or geographic boundaries, and he offers a solution to a perplexing problem. It is tempting to see environmental problems as bounded within jurisdictions, as one might refer to "The pollution problem in Georgia," for example. This viewpoint encourages an idea that politics and political institutions preexist problems and limit problems within political boundaries. But even a cursory examination of the real world dispels this image; most environmental problems—especially second- and third-generation problems—cross political boundaries, whether local municipal boundaries, state boundaries, or international boundaries, and the origins of public awareness of an environmental problem are often clouded in confusion. Trying to match problems to jurisdictions, in the face of rapid technological and social change, seems futile.

As an alternative, Dewey hypothesizes, "[t]hose indirectly and seriously affected for good or for evil form a group distinctive enough to require recognition and a name. The name selected is 'The Public'" (Dewey, 1927, p. 257). This means that problems as felt by citizens exist in a logically prior status regarding political institutions, but in itself the public is "unorganized and formless" (p. 277). A public thus takes on its moral and political form through the emergence of representative spokespersons who, in turn, form institutions that identify and attempt to manage consequences of past and current actions.

So for Dewey, problems are not initially bounded within political jurisdictions or associated with particular institutions. He recognizes that the concern of citizens can create movements that are responsive to aspects of problems, movements that stretch across political boundaries. Observers of evolving environmental problems have noted that the environmental problems faced today almost never "fit" within any given political jurisdiction (Hajer, 2003). Rivers suffering from pollution flow across state and international boundaries; pollution from tall smokestacks blows from one region and upsets natural systems in other regions. Climate change and other third-generation problems outpace the development of effective global institutions to manage them.

While this sounds like a political catastrophe, empirical observation tells us that there are many instances in which, despite inchoate beginnings, despite scientific uncertainty, and even despite the obvious inadequacy of existing institutions to address problems playing out on larger and larger scales, the efforts of some communities have been surprisingly successful (Wondolleck and Yaffee, 2000; Blomquist and Ingram, 2003; Hajer, 2003). As it turns out, complex problems often require complex institutional arrangements and frequently lead to the emergence of polycentric organizational structures.

If the world is seen from a pragmatic, problem-oriented perspective, then our exploration can truly begin with concern for consequences. As Hajer so aptly puts it, responses to environmental problems often succeed by creating "policy without polity." The phrase, while apropos, is of course an exaggeration; he recognizes the ultimate role of sovereign states and legitimate jurisdictions, yet he sees solutions as emerging from informal and ad hoc organizations that, at their best, can influence governments and contribute to solutions that could not be conceived or implemented within any one jurisdiction. Hajer's point, then, is not that governments are unimportant, but that successful response to complex environmental problems begins with public discourse and the development of a public sphere of informal and advisory organizations. These organizations, formed of individuals who care deeply about problems, can be more effective than regular political channels (Ostrom, 2007, 2010). Such approaches have been quite successful in water agreements across jurisdictions, and there are excellent examples that can provide some guidance (Wondolleck and Yaffee, 2000; Blomquist and Ingram, 2003; Freeman, 2010).

This recognition suggests an important hypothesis about the transition from "messes," in Ackoff's sense ("wicked problems," in the words of Rittel and Weber), to problems with enough structure to allow analysis and the evolution of partial solutions in response to the messy situation. A hypothesis is that the development of institutions and organizations is an essential step in moving forward with a process to address wicked problems. One has not, in other words, successfully posed a problem until the inchoate concerns highlighted by Dewey result in organizations and institutions forming goals and missions that define the problem by articulating possible responses.

Once this step is well under way, it will be possible to see that a public, understood as a population concerned about possible consequences, can develop into a "community"—a group of individuals who form associations and institutions to address common concerns. The emphasis will not be on a single jurisdiction as the locus of problems; rather, the emphasis should be on creating polycentric organizational structures. Informal arrangements emerge at different scales and across differing jurisdictions. One way to explain this point would be to say that the inchoate concerns of a Deweyan public can lead to a clear articulation of an addressable problem, as individuals form groups that set certain key "boundaries" to their concerns by organizing and institutionalizing their activities. The importance of this work at the boundaries of problems is sufficient to justify a subsection of its own.

5.3.3. Problems, Institutions, and Boundaries: The Problem of Problem Formulation

If most environmental problems are wicked problems or have wicked aspects, then attention to problem formulation is clearly indicated; unfortunately, this part of the process of environmental policy formation remains as an unexamined element of decision science. Decision scientists have known for decades that formal models cannot tell one how to formulate a problem, and yet a virtual void is faced in problem formulation, which is the first step in intelligently addressing a problem.

My reaction to the void has been to insist on forms of rationality that do not depend on algorithms and computations of ideal solutions; and this brings us to the crux of the problem with problem formulation. The goal of much of decision science is to formalize a process that cannot be formalized. In reality, problems emerge and are sometimes dealt with (usually partially), not as a result of a computational model but through the institutional arrangements that self-organize within a public as it becomes a community, making it able to act cooperatively toward limited ends consistent with many different values and interests. But there will also be competition among institutions and organizations that emerge in particular situations to serve particular interests in the process of problem formulation.

This process of developing a Deweyan public to the point where it can identify actionable problems is thus essentially context-specific. There are no one-size-fits-all solutions, and there is not a single method that alone will generate a correct answer if given enough facts. This process is visible, if one looks carefully, in particular contexts where a community is developing, as institutions and groups strive to control the policy agenda. This process cannot be routinized, nor can it be guided by optimization models, nor can it be understood independently of the context in which it unfolds. The process is a richly social activity, highly conditioned by the special features of each unique situation.[1]

Getting on the road to better problem formulation requires dealing with problems of scale, which, in turn, raises issues as to how boundaries around problems are shaped. By following Leopold's simile of thinking like a mountain as a guide to understanding the human–nature interactions in particular places, one comes again and again to the central issue of time, of the many scales and levels on which one can understand the natural world, and of the open-ended nature of environmental monitoring and management. Many environmental problems represent clashes between the temporal scales of human activities and nature's processes. Hierarchy theory (HT) can provide some useful vocabulary and helpful models for talking about the multiscalar world, even though HT is only a formalism and does not by itself resolve questions of scale. When HT is used to represent spatial relations in actual situations, analysts need to make many choices.

The hypothesis offered in the previous subsection, that proceeding toward better problem formulation in wicked situations requires the development of institutions and organizations, leads to an important realization: a problem is generally not well formulated in the absence of vital institutions that compete and cooperate in pursuit of goals that may or may not be widely shared. Acting on this hypothesis brings together, under the heading of the "problem of problem formulation," three main themes of this book:

1. Wicked problems and their open-endedness
2. Problems of scale and system boundaries
3. The importance of work that must take place across "boundaries" in order for publics to form and to move toward more effective problem formulations

I will begin with the third theme and explore all three themes from the viewpoint of "boundary work." Fortunately, a significant and growing literature is discussing the importance of boundary work. Originally, the idea of such work arose when sociologists addressed the "demarcation problem": how, in practice, is science separated from pseudoscience and fraud? This work originally referred to the creation of either concrete or abstract "objects" that serve to link individuals and groups across important boundaries. Such a boundary object might be a model of a watershed that is shared, discussed, and criticized by both scientists and policy makers. Or the object might be an institute for the scientific and policy study of a problem. The original definition, given by Star and Griesemer, is as follows:

> Boundary objects are objects which are both plastic enough to adapt to local needs and constraints of the several parties employing them, yet robust enough to maintain a common identity across sites. They are weakly structured in common use, and become strongly structured in individual-site use.

They may be abstract or concrete. They have different meanings in different social worlds but their structure is common enough to more than one world to make them recognizable means of translation. The creation and management of boundary objects is key in developing and maintaining coherence across intersecting social worlds. (Star and Griesemer, 1989)

Boundary objects will become important again in chapter 9, when the communal nature of some forms of social learning is explored. Careful attention to boundary work has led science-policy experts, especially those interested in how sciences maintain their boundaries from questionable and fraudulent pseudosciences, to growing interest in organizations that operate on both sides of a divide, and especially, with those that help demarcate responsibilities of scientists and policy makers in advisory relationships affecting regulatory policy.

This idea of a separation, but with a permeable membrane between the science community and the policy community, can guide the search for "appropriate" processes by which to decide environmental problems. Boundary organizations, it has been theorized and to some degree empirically corroborated, can improve communication and build trust on both sides of the science-policy divide (Guston, 2001). Further, it is possible to expand these concepts of boundary objects and boundary organizations to cover many situations where distinctions are drawn and boundaries are set; in this expanded context, the boundary work spanning science and policy is merely one type of such work.

The idea of boundary organizations spanning science and policy communities has given rise to another application of the idea of boundary objects and boundary organizations: the ideas can be applied to processes that function across geographical boundaries in negotiations to protect cross-boundary resources such as large aquifers (Blatter and Ingram, 2001). Research on such conflicts has shown that, despite claims by experts that allocation should go to highest uses, the most successful cross-border water management schemes have emerged through development of ad hoc and advisory groups and institutions. This, of course, is what Adapt has been saying all along. These groups begin working on relatively neutral aspects of the problem, such as agreeing to keep comparable records and establish monitoring standards. These steps create ongoing processes, and parallel scientific and policy organizations form on either side of the border. Iterative interactions tend to create trust and encourage possibilities of developing "boundary objects" such as a shared model of the dynamics driving concerns on both sides of a political border. This insight can be stated as heuristic 6.

Heuristic 6

The boundary institutions heuristic. Construct boundary objects and institutions that encourage a community of interests. For most complex problems today, there is no "spatially appropriate" institution or jurisdiction available; let emergent public voices articulate problems and bound appropriate institutions.

5.4. Public Participation: Dynamic Evaluation and Sustainability

I have not entirely lost the thread of the story that highlights sustainability and how contemporary, diverse societies might find a path toward sustainability, understood as normative sustainability. Sustainability for a community can emerge from a public process that works together to protect highly valued opportunities that are associated with aspects of a local place that are important to its denizens.

Having given up the fool's errand of calculating in advance an outcome that is perfectly sustainable, given some theory of environmental value, I turn my attention instead to the public in search of sustainability values. Heuristic proceduralism relies on a process that improves the likelihood of achieving cooperation in the pursuit of partial, but broadly acceptable, outcomes. My approach to evaluating change is to weigh possible development paths for a community according to their effectiveness in protecting the most important opportunities that are manifest in the unfolding dynamics of the ecological systems constituting the "place" the community occupies.

When someone insists that a given physical process is important to monitor, one can infer that the process is perceived to support a value of that individual. The judgment that a given process is important enough to monitor suggests treating it as an indicator—one that may reflect values expressed in multiple ways and perhaps even multiple values. I thereby bypass ideological arguments about why something is valuable and instead pursue identification of stuff (that is, key elements and processes) that must be saved to support opportunities to enjoy aspects of natural systems, however the individuals involved would explain that value.

5.4.1. Choosing Indicators as an Expression of Community Values

My new approach to evaluating change is thus a deliberation-based process by which a public learns to protect key natural processes because they support opportunities that, in turn, support a variety of values. The search for sustainability thus incorporates public deliberation based on values but

Figure 5.1. Deliberating toward sustainability: Adapt recommends directing discourse toward the identification of indicators that will be considered important enough to monitor if the variable in question supports valued aspects of the ecosystem and processes.

expresses them in the form of indicators thought important by individuals and groups that support diverse values. If a community can identify a list of key indicators that support values they revere, it may not matter that they would explain those values in different language or offer different supports for them. This approach allows and encourages the development of coalitions behind the protection of key variables by deflecting questions away from ideology toward choice of variables to monitor. This process represents an attempt to operationalize "the public interest" for a given community.

By concentrating on the process of choosing indicators, a community can direct attention to what is considered important to save, which can be less divisive than arguing about which elements of a natural system have value or of what type that value is. (See figure 5.1.) In many situations, cooperation among disparate interests in a community or collaboration across boundaries to solve larger problems begins with agreements to monitor some set of variables or to begin seeking scientific advice.

But choosing a set of measurable and informative variables as indicators is no easy task: it is hard work and requires deliberation, because choosing what to measure will greatly affect the subsequent stages when more substantive actions become necessary. Choosing indicators must therefore be a matter for public deliberation among all stakeholders, constituents, and affected parties.

As important as the task of choosing indicators is, one cannot seriously address that task at the general level of a philosophy of management. As just argued, the choice of indicators should be guided by deliberation of the community in question. There will be no one-size-fits-all criteria for success—each community must express its sense of place in the process of choosing indicators that highlight processes important to the its traditional and evolving values (Innes and Booher, 2000, 2010).

What can be done is to provide some guidance as to the types of indicators that may prove useful by discussing an example. A few years ago, the city of Atlanta, inspired by successful programs in Seattle and other cities, was inspired by community groups and city employees to undertake a "Sustainable Atlanta" exercise. At the first meeting, with over a hundred people in attendance, it was announced that the audience was to come to the next meeting with a list of indicators they thought would be useful markers of the city's sustainability. The next week, on request, about 40 indicators were proposed. Before the end of that second meeting, a call went out for more indicators. In the end, the participants were responding to a list of 150 indicators. Even a pluralist realizes one cannot manage for 150 variables, so what is a reasonable number of indicators for a community to track? Again, there's no cookie-cutter answer, but I can offer a few helpful observations.

Having rejected the goal of optimizing one or a very few variables in part 1, and now having rejected a list of 150 as too long, we are still without much guidance. Some number of criteria, in the form of indicators and standards, should be identified by deliberation within particular communities. How these criteria are organized and related to others across multiple scales is probably more important than the number of indicators themselves.

Perhaps providing a hypothetical example would help. Faced with a plethora of proposed indicators, one can suggest that, in addition to primary indicators originally proposed, it is useful to try to consolidate the list by creating what has been called "synoptic indicators."[2] Consider an example: perhaps the city of Atlanta and the Atlanta Regional Commission should create for the whole area, as well as for smaller municipalities within the region, a program to track and publicize trends in the protection and loss of pervious surfaces (meaning soil areas uncovered by roads and parking lots). One important reason to consider this as one useful indicator is that it can be measured quantitatively and accurately, using existing Landsat satellite data that is routinely collected. In general, precision and avoidance of huge costs for monitoring would affect the choice of which variables can make useful indicators.

Notice that few residents or experts would list pervious surfaces as one of their important *values*, and yet trends in the amount of such surfaces might make a very useful indicator because, since it reflects social values indirectly, it can be the basis for building coalitions favoring activities that positively affect the currently negative trend toward more concrete and other impenetrable surfaces. Impervious surfaces are responsible for much of the runoff that fouls streams and also robs aquifers of recharge water that instead flows in torrents to the sea. So, paying attention to the easily measured variable of trends in pervious surface loss can be helpful to advocates of clean water and to advocates of protecting water quantity. Similarly, we know impervious surfaces are bad for wildlife and for biodiversity more generally, so wildlife advocates will concur that this indicator is a useful one, and they will raise alarms when losses continue unabated or accelerate. Likewise, advocates of smart growth will argue that this variable is relevant to their concerns and that decreasing sprawl will protect natural systems even as it creates more close-knit and livable communities.

Now, surely this single variable and criteria associated with it cannot do all the work of measuring success at becoming sustainable; but it could serve as an easily tracked variable that can set off alarms when trends toward greater losses are observed. Here, it is useful to remember that it is easier to define and recognize unsustainable actions and processes than truly sustainable ones. Further, the indicator can unify individuals and groups that can form coalitions to support actions directed at affecting the variables chosen.

Tracking pervious surfaces is one example of an indicator that combines ease in measurement even as it maintains an intuitive relationship to a number of social values. One can, referring to these two characteristics, suggest other examples. With respect to aquatic systems, turbidity of the water column, which typically measures the amount of algae and other sources of turbidity associated with overnutrification of a water body, similarly is very easily measured. When increasing rapidly, this measure indicates negative trends likely to affect values such as fish productivity (affected by turbidity that reduces underwater vegetation), aesthetic appreciation, and recreational opportunities. As another example, with respect to ocean systems in tropical waters, extent of coral reefs might make a useful synoptic indicator, given the many ways in which these diverse systems serve important values that nearby communities should care a great deal about. By now, it should be clear that choice of indicators will be highly idiosyncratic and determined more by local values and characteristics of local ecosystems than by an abstract definition.

To highlight the key role of choosing indicators, I introduce heuristic 7.

Heuristic 7

The sustainable indicator heuristic. Evaluate development paths according to their likelihood of achieving the public interest conceived as a widely acceptable balance in protecting important variables—a suite of indicators deemed important enough to measure and monitor as sustainability indicators. Arguments that a given indicator is important and should be monitored express values of participants in the process.

Once the measurable indicators of importance are chosen (tentatively, of course), there will still be disagreement about both the comparative importance of the indicators and the pace of change that is acceptable with respect to chosen variables. Public debates, however, can be directed at determining the proper weighting of indicators and at choosing policies that will protect key variables.

5.4.2. Public Participation: Sick or Healthy?

In this section I concentrate on characteristics of healthy and productive public participation, recognizing that our heavy reliance on procedural criteria places strong demands on public participation. That this is not merely a dream or a remote change far in the future is shown by a brief history and evaluation of processes for public participation, beginning with the National Environmental Policy Act's (NEPA) edict that projects and policies of a certain complexity must go through a process of public readings and hearings. Since the passage of NEPA in 1969, agencies and other actors have been required to provide opportunities for public involvement in decisions with significant impacts. At first, this requirement was viewed skeptically by managers and agency heads alike, as they feared that public involvement would raise issues and complicate the process. Early responses by officials to NEPA were either to design projects so that they fall under the scale at which an environmental impact statement is required or, if that was impossible, to comply strictly with the letter of the law by holding public hearings for a project only after a full plan was completed and implementation was imminent.

Shepherd and Bowler (1997) showed that this type of participation does not, as managers hoped, reduce public opposition to projects. Indeed, when the public is faced with a fait accompli, in the form of a plan so fully specified that the only option is to accept it or reject it, they often oppose such projects, participation then becomes polarized, and finally members of the public turn litigious.

Another weakness of early attempts to comply with NEPA was a tendency to view participation as a single event designed to "get the project approved." Public involvement, it has been learned, is more effective when engaged from initial planning through the implementation stages. Learning how to effectively involve the public in complex decisions is a learning process, because every problem has its own, uniquely wicked aspects. It is clear that "rubber-stamp hearings," sometimes referred to as a DAD strategy ("Decide, Announce, and Defend"), simply do not work. Since the early days of NEPA, steady progress has been made in achieving better understanding of which processes work, and considerable progress has also been made in implementing better processes.

In response to these concerns about public participation processes, Shepherd and Bowler suggest four critical factors in "developing nurturing and trustful relationships by means of democratic and honest negotiations with all stakeholder groups." Their criteria for successful participation processes can provide us with important insights for the design of such processes. They refer to their four critical factors as (1) democracy, (2) suitability, (3) conflict resolution, and (4) improved planning (Shepherd and Bowler, 1997, pp. 728–29):

1. *Democracy.* Public participation in environmental decision making aspires to an ideal; democratic involvement helps ensure that all affected voices are heard. In the best cases, when members of the public are involved in a project from the beginning and believe their viewpoints and suggestions are taken seriously, they gain both a sense of ownership of the actions and also a sense of political responsibility for having taken them. This process feeds social learning; further, even small agreements in tense situations can break the ice and increase communication and trust, which can pay dividends in future cooperation (Shepherd and Bowler, 1997; Sabatier et al., 2005).

   It would, of course, be foolish to suggest that all, or most, public participation processes achieve or even approach the ideal that every affected voice is heard and that every participant has equal input in final decisions. In the best of cases, to quote Shepherd and Bowler, "citizen involvement helps to guard the public interest" (p. 728). Agreeing with Shepherd and

Bowler, one can argue that the more democratic a process of participation is, the more one can privilege its decisions.

2. *Suitability*. The suitability of a process, as Shepherd and Bowler explain, also can affect whether members of the public adopt a sense of ownership of projects and policies. They suggest that the more suitable a process, the more likely the outcome will be embraced and not fought in courts. "Suitable" in this context can be treated as synonymous with "appropriate," thereby connecting characteristics of public processes to process rationality by definition.

3. *Conflict resolution*. As one thinks about environmental policy development and implementation in the posttransition mode favored here, the emphasis will be much less on analysis and computation and more on conflict resolution and creative problem solving. Entering deliberative processes with an open mind is critical. When some participants think they know the correct answer, or believe that application of their theory or method will generate right answers, they rule out options a priori, without fully looking at alternatives. If, on the other hand, the conversation starts with everyone on an equal footing, asserting their interests, it is possible to apply techniques of conflict resolution and mediation to move toward policies that are acceptable to more participants. If participants' attention can be displaced from their favorite solutions and monistic theories about what is "good" to a discussion of which alternatives will protect those processes that support the widest possible range of values, then progress might occur.

4. *Improved planning*. Shepherd and Bowler also see it as critical that public participation be designed and carried out with an eye to improved planning. Not only are citizen participants a source of expertise, their varied viewpoints also provide alternative perspectives that must be taken into account in planning for the future. When the time frame is stretched to multiple generations—planning for sustainability—present participants are the source of information about which opportunities and values are sufficiently important that they must be protected for the community in question to retain continuity with the past.

Since it is impossible to know in any detail what people of the future will want, the present citizen participants will be the main source of reasons and justifications for identifying sustainability goals, goals that will protect the key values that give meaning to the community's multigenerational identity. The opportunities that are essential to protect in order to maintain community continuity over time must be identified democratically through a process by which current citizens advocate for the importance of certain ecological and other physical dynamics by proposing and defending effective indicators. Choosing an indicator involves deciding which dynamics are important enough to be protected for future generations and thus involves a commitment—a moral performative—to protect the core values of the community.

While some public participation processes fail because they still operate on the minimal practices that Shepherd and Bowler criticize, much progress has been made in understanding what makes a public deliberative process effective; there are many examples where agencies have engaged the public constructively. These new approaches provide protocols that, if followed, encourage cooperation and social learning—learning that is profound enough to question and revise goals and objectives as well as scientific understandings. One such process model, which was developed to replace the overly scientistic approach to risk assessment and management that dominated the field until the 1990s in the US, captures most of the important points I have, in this subsection and elsewhere, made about high-quality participation in decision making.

There is no foolproof way—indeed, it is extremely difficult—to create an effective, fair public participation process. The diagram in figure 5.2 provides a template for designing successful public involvement in complex problems. The panel assembled by the National Research Council got it right as an abstract representation of what needs to happen. The hard part, of course, is to apply this abstract design to particular, difficult decision situations. This diagram reflects what would face an adaptive manager who realistically wants to address a wicked problem, recognizing the cross-cutting interests that will be affected by anything that's done. It provides a path forward, no matter how many setbacks occur. The strategy, which must be developed in the particular contexts, builds from the bottom up, a process that could support better environmental decisions, even in difficult situations. By shifting to an ideal of rationality that pursues not final answers but instead appropriate processes, it becomes possible for public discourse to focus on problem articulation, on finding partial solutions, and on forging compromises that balance competing values.

5.4.3. Deliberation and Social Learning

The European political scientist Manuel Arias-Maldonado has recently questioned the role of deliberation in achieving an "open" understanding of sustainability (Arias-Maldonado, 2012). Earlier in this book, in section 1.2, it was noted that his preference for open approaches in search of sustainable policies parallels my advocacy of Adapt over Optim. He sees closed and open approaches as "meta-theories about the way in which [sustain-

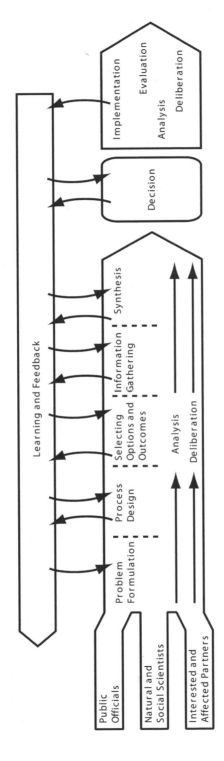

Figure 5.2. An iterative, public involvement model: The National Research Council, when asked to sketch out a program of public involvement in risk assessment, provided this iterative model that, if followed, would greatly improve public participation (National Research Council, 1996).

ability] should be approached" (p. 108), which corresponds to the strategies of Optim and Adapt. He describes closed approaches as constrained at the outset of the inquiry by "ideological coherence" or "technical feasibility," which fixes the goals of sustainability. These goals, once objectified, determine the outcomes that will be sought, leading "de facto to the abolition of both politics (understood as the intersubjective realm for deliberation and negotiation) and society (as the practical realm wherein deliberation and negotiations take place in a non-formal way)" (p. 108). For Arias-Maldonado, open approaches, by contrast, eschew initial identification of the key aspects of a sustainable society, leaving it to the ongoing processes of political and social development to spell out such a society's features. His open version of sustainability "does not determine in advance the content of a sustainable society, which remains uncertain and will eventually emerge as the outcome of a liberal-pluralist society, wherein a permanent controversy about the good life is discussed in the public sphere and played in the practical realm." Further, "[t]he open view is dynamic and prefers a liberal-democratic setting—complemented by governance elements—as the sociopolitical means by which different ideas and practices of sustainability can be tested in a pluralist context" (Arias-Maldonado, p. 113).

Despite the initial similarity of Arias-Maldonado's open sustainability approaches to the adaptive approach I have outlined, and despite his mention of "governance elements"—which presumably refer to public processes allowing citizens to affect policies in ways beyond simply voting—he criticizes deliberative methods as politically inadequate to pursue sustainability. In this section, I compare the limited role that Arias-Maldonado assigns to deliberative democracy with the more robust form advocated here, and I also show that advances in the psychology of behavioral change support a more robust process such as the one developed here.

Arias-Maldonado mentions five reasons why advocates of an open, strong sustainability are drawn to deliberation as a means to achieve their vision: (1) deliberative contexts encourage the emergence of environmental ("green") values; (2) deliberative democracy can bring more, hitherto excluded, individuals and groups into the process; (3) deliberative approaches provide "the best institutional arrangement for developing ecological citizenship"; (4) deliberation results in the interaction between citizen-participants and experts, which can encourage application of science and technology within a normative context set by public decision making; and (5) inclusion and deliberation in decision making can contribute to "more legitimate and efficient decisions on sustainability" (Arias-Maldonado, 2012, pp. 141–43).

One might expect that, having provided this impressive list of advantages of deliberative approach, Arias-Maldonado would proceed to roundly endorse deliberative processes as highly important in seeking sustainable policies. Nevertheless, immediately after listing these considerable advantages, he explains why deliberation is insufficient to guide a society or community to sustainability. He offers two arguments to show that introducing deliberative institutions may, at best, offer a "supplement" to traditional economic and political processes. First, he questions how successful deliberation will be in developing more sustainable attitudes (a question I will address in chapters 6 and 7); second, he fears that developing processes alternative to usual economic and political ones may bias outcomes as citizens pressure others to conform with their ideas.

Arias-Maldonado's alternative to the emphasis on processes and adaptive procedures, such as are developed in this book, rests on two supporting claims:

1. The main arena for searching for open sustainability should be the existing, democratic governmental structures of state societies, including representative (rather than mainly participative) institutions.
2. Sustainability advocates should not be too optimistic about the power of participation and deliberation to create a dynamic that will shift attitudes and change the behavior of participants toward more sustainable practices.

With respect to claim 2, Arias-Maldonado states, "What we need most, in short, are enlightened people rather than deliberative groups" (pp. 145–46). Note that, at each point at which he raises skeptical concerns about the contribution of a participative, deliberative approach, he shifts into the subjunctive mode. For example, "But it *may* be that an institutionalized deliberation is not the right setting for generating such wisdom, although it can perform a modest, supplementary function within a representative democracy and a liberal society" (my emphasis, p. 144). Many other passages in which Arias-Maldonado questions the efficacy of deliberation are laced with "mays" and "cans." Such subjunctive claims, at most, indicate that, in some or many cases, deliberation may fail to create transformed preferences. We should not be surprised by this: we know that many deliberative processes fail. The question, addressed in chapter 6 and the remainder of this book, however, is whether (assuming good faith on the part of participants and best procedural practices represented by heuristics) communities can learn to be sustainable. Given appropriate processes and good faith on the part of participants, social learning through deliberation can advance the cause of sustainability. I have set out to provide a set of heuristics that can guide a community forward toward sustainability; these heuristics, taken together, empower adaptive processes beyond what Arias-Maldonado attributes to them.

Arias-Maldonado's claim 1 is mainly supported on the basis of his claim 2, so our final verdict on claim 1 must await the complete development of procedural approaches and heuristics as they unfold in this book. Nevertheless, we can respond to claim 1 by noting that the process being developed here—and that suggested by other advocates of adaptive management—does not envision replacing normal governmental agencies and functions. Advocacy for adaptive management is by no means a desire for a coup d'état. Many, perhaps most, such processes are intended to develop proposals and to lead from an advisory capacity. The goal is to create an integrated structure that considers policy and actions from a more environmentally concerned point of view.

For example, stating that a number of environmentalists have already accepted that the role of deliberation will be to contribute to ordinary government processes, Arias-Maldonado approvingly quotes Albert Weale, who says, "Representation, not participation, is the key problem of environmental policy making" (Weale, 2009, p. 56; Arias-Maldonado, p. 149). Calling a "retreat" the recognition that normal representative functions of government must remain in place during deliberative processes, he says, "this retreat goes hand in hand with a rediscovery of the state—a green state—as an enforcer of sustainability" (Arias-Maldonado, 149).

Now it is clear that Arias-Maldonado, by exaggerating the role attributed to deliberative processes by its advocates, falsely suggests that the advocates of deliberative processes seek the role of final decision maker (which they seldom do). If one understands deliberation from the beginning as complementary to normal governmental processes and if one recognizes that most adaptive processes provide advice and proposals for improved environmental and sustainable policies, then one can develop adaptive, deliberative processes to generate better decisions than can relying only on existing governmental institutions.

The importance of Arias-Maldonado's pervasive use of the subjunctive voice when pointing out weaknesses of deliberation is now clear. By correctly pointing out that simply introducing the possibility of deliberation into a system of governance may not lead to sustainability, he has merely posed the problem for this book. Naturally, while some public processes fail, others succeed; the interesting empirical question is: is it possible to tell the difference, and thereby provide guidance to those engaged in public dialogue and deliberation? I thus grant to Arias-Maldonado that adaptive collaborative processes will not replace existing, representative governmental structures, since I have never questioned that the pursuit of sustainability must involve existing and, one hopes, reformed democratic governmental institutions.

Consider common situations in which an important resource that spans existing governmental jurisdictions is being degraded—as, for example, when a river separating two states or nations is being polluted from both sides, creating problems that span political boundaries (Hajer, 2003). Typically, in such cases, governments on opposite sides of the river blame the other party for both the pollution and the failure to clean it up. When left to their own devices, such standoffs can last indefinitely as a crucial resource continues to degrade. But there are examples where local people on both sides of the river join together to form a deliberative group; often they begin with neutral activities like instituting common monitoring conventions for both sides of the river. In many of these examples, there is evidence that such informal, extragovernmental processes can set overarching goals and exert pressure on both governments to change policies and contribute to the overall goals by "supplementing" normal governmental functions (Blomquist and Ingram, 2003).

Therefore, I put little weight on Arias-Maldanado's minimizing the contribution of deliberation and collaboration to the development of sustainable communities. One need not see deliberation and adaptive collaborative management processes as competitive with, or alternatives to, existing, state-based political institutions. All these processes exist in the broader realm, that of governance, in which informal and formal procedures interact.

One can also raise questions about Arias-Maldonado's reasoning (based on his claim 2) that deliberation is unlikely to change preferences. He says, "There is no guarantee that ecological values will be embraced as a result of free and equal deliberations. . . . [I]f we pay attention to the current knowledge on practical reasoning and political motivation, we see that it is not reasonable to expect that citizens will dramatically change their preferences through deliberation" (Arias-Maldonado, 2012, p. 144). Again, Arias-Maldonado overstates his case. One need not believe in miraculous and "dramatic" changes in preferences to believe that better policies and sustainable policies can emerge from deliberative processes. In fact, preferences and changes in preferences may occur slowly, yet preference change is not the only benefit of an ongoing adaptive, collaborative process. A major difference between my approach and that of Arias-Maldonado is my introduction of adaptive management processes that emphasize action and experimentation and thus relate more closely to behavior. Adaptive management is not limited to passive information provision; it involves activities and social learning that may generate experimental actions, such as pilot projects. So my strongest disagreement with Arias-Maldonado is that he considers "deliberation" to be passive and not part of an active, learning-by-doing process.

The psychologist Thomas Heberlein (2012) has recently shown that behavioral change often will not result from preference or attitudinal change and that changing behaviors usually occur as a result of situational factors affecting the decisions individuals face, including especially structural features of that situation that affect incentives. This suggests that we need to envision a more complex, dynamic, and activist process to understand how social movements such as the movement toward sustainability advance.

A more detailed examination of Heberlein's important work will differentiate my approach from that of Arias-Maldonado and clarify some cross-disciplinary confusions. Heberlein, as a psychologist, examines the role of attitudes, values, and behavioral norms, and he develops his argument using somewhat different terminology than is used in this book, though the points he makes are of great importance in our search for sustainability.

Heberlein focuses first on "attitudes," which are pervasive in human life and which he describes as beliefs "tied to a value," such that the belief "says or implies that something is better than something else." Emotion (what psychologists call "affect") gives force to attitudes (Heberlein, 2012, p. 16). Attitudes always have an object, and in this respect, attitudes are relatively stable across time but, unlike economists who believe that preferences or attitudes are fixed, psychologists believe attitudes do change—slowly and only under the right circumstances. When one first develops an attitude toward an object, one's information base is typically small and subject to change, but as one has more contact with the object, and as emotional aspects build, attitudes can become quite stable. What stands out in Heberlein's treatment is the way he relates attitudes and the information base supporting them to behavior. He argues persuasively that attitudes, by themselves, are usually not sufficient to lead to appropriate behaviors and that changes in attitude, by themselves, are unlikely to lead to behavioral change. He hastens to add, however, that attempts to change behavior are much more successful when the behavior is consistent with the person's attitudes than when behavioral change goes against those attitudes.

Heberlein therefore criticizes most attempts to change behaviors that have detrimental environmental effects, referring to these attempts as "cognitive fixes." Heberlein argues that the cognitive fix, the provision of information and attempts to educate the public, is almost never successful. He provides an example: many rivers flood, causing great damage to people and their structures built on floodplains. He describes how the state of Kansas mounted an information campaign, complete with updated floodplain maps and messages in the media, trying to convince people not to build on the floodplain of a river that flooded seldom but regularly. The effort failed, because residents, looking at a floodplain that is dry almost every year, doubted and ignored the maps and went ahead and built on floodplains, leading to serious damage when the eventual high-water spring flood occurred. The behavior of building on the floodplain was unaffected by information and education. However, when financial institutions, stung by necessary repossessions and losses due to flooding, required floodplain information on mortgage applications and refused to write mortgages to cover floodplain development without expensive insurance, behavior changed. Building in the floodplain virtually ended (Heberlein, p. 6).

Heberlein's key point is that behavioral change usually results from what he calls a "structural fix," which is a "change in the social environment that influences what people do" (p. 6). Often, this involves a change in incentives; the incentives operate within a system in which attitudes remain important but not determining. Incentives work more successfully if they are consonant with existing attitudes, so attitudes are relevant. Heberlein argues, however, that attempting a cognitive fix without structural changes that affect incentives seldom changes attitudes, and thus attempts to change behavior simply by changing attitudes is almost always ineffective.

What is more effective, he states, is to understand attitudes and attempt to work in consonance with them by changing the incentive structure so that individuals will change their behavior. If the incentives move actors toward behaviors that are more environmentally sound, and if those behaviors come to be expected—and if failures to engage in those behaviors are censured in some way—a "behavioral norm" can emerge. Such norms, which are stabilized when they are sufficiently followed, affect behaviors, since failures to act in accord with a norm can lead to sanctions ranging from a "dirty look" to legal prosecution.

To understand Heberlein's implications for this book—and also to answer some of Arias-Maldonado's criticisms of deliberative approaches—it is necessary to clarify cross-disciplinary terminology. In this book, I have used the term "value" as a general term to refer to people's preferences, evaluative attitudes, norms, and also deeply held principles, as is common in my field, which is philosophy. Heberlein, a psychologist, by contrast, separates attitudes from values by saying that the former always have an object, whereas the latter do not (Heberlein, p. 15). Values, so understood, are much more stable than attitudes, especially if they connect to a person's or a group's identity. Heberlein thinks it is quite difficult to change values—they change, but over years and decades—while attitudes, even ones based on experience and a strong information base, can be changed, given the right approach. Since environmentalists and adaptive managers are concerned with both values and attitudes and with the effects they have on behavior, I have learned from Heberlein's approach that, using his more formal terminology, distinguishes among these value concepts, examining attitudes, values, and behavioral norms and assigning a special connection between norms and behaviors because norms are regularities in people's actual behaviors (Heberlein, p. 92). Norms regulate behavior, and in some cases, a norm is simply a pattern in behavior. But, in the cases relevant here, norms can come with sanctions and rewards that, in turn, provide incentives to act. So if we are concerned with altering the behavior of a social group by, say, encouraging the development of norms protective of the environment, then a structural change in the situation, such as instituting negative or positive incentives, or even both, is the most effective means to do so.

Because much of the research and writing in this book was completed before I read Heberlein's book, I have not fully adopted his formal distinctions in all situations. His finer distinctions, however, not only are consistent with my use of the broad categories of value and facts but actually support and strengthen the adaptive management approach and help me to provide a clear explanation of how collaborative processes, when well designed and executed, can lead to improved policies and to trends toward more sustainable living.

As noted above, Arias-Maldanado apparently assumes that "deliberation" refers to sharing of information and discussion, and he concludes that deliberation will not be sufficient to support behaviors on a path toward sustainability. Without disagreeing with him regarding the necessity of maintaining regular democratic governmental processes, I also believe that simple provision of information and holding passive discussions are insufficient, as Heberlein shows. I disagree, however, with Arias-Maldonado's understanding of what can be included under the term "deliberation." He seems to accept a narrow conception of deliberation that includes mainly information and discussion. He writes, "In the end, it is information that matters most: citizens who engage themselves in experimental deliberative arenas are firstly shaken by the new information they receive—it is only subsequently that the exchange of arguments with other people will benefit or, as we have seen, distort the collective decision. What we need most, in short, are enlightened people rather than deliberative groups" (pp. 145–46).

Given this quite limited role assigned to deliberation, it is not surprising that Arias-Maldonado can easily make the case that deliberation alone is unlikely to change behavior drastically in the direction of sustainability.

If, however, one understands deliberation within an adaptive process that includes discussion and debate but also embraces broad representation by scientists and experts, experimentation, and especially, the second-order learning involved in following heuristics in order to make adaptive processes more effective, then that process will be able to challenge norms that are unsustainable. It will then encourage the development of norms that might lead to sustainability.

More specifically, informal groups can reach agreements that affect its members' behavior by developing rules that everyone agrees to follow and by creating sanctions to enforce them, as has been advocated by Ostrom (2007) and colleagues. Likewise, citizen groups can propose positive incentives, such as tax breaks for behaviors that will be protective of the community's goals, and can encourage formal governmental institutions to put these incentives in place. True deliberation can lead members to propose positive incentives and rules for enforcement that will affect the structural situation faced by members and their cohorts. These proposals and the incentives they create may require enactment by governmental agencies and the various levels of government. So deliberation cannot replace normal governmental institutions but rather can provide recommendations and proposals that have the support of informal organizations. Reflective, adaptive management thus offers far more effective processes than ones based simply on providing information. Because he undersells the potential of deliberative processes, Arias-Maldonado falls back on the governmental institutions that have so obviously failed in their occasional attempts to achieve sustainability.

The key point here is that once an adaptive management process becomes self-conscious and refers to heuristics to guide its process, and once its practitioners learn how to develop and introduce incentive systems that will actually affect behavior as suggested by Heberlein, one can expect much more of deliberation than Arias-Maldonado does. Self-conscious adaptive management can incorporate the insights of psychologists and other social scientists to improve the process and make deliberation—now enhanced by heuristics and supplemented by incentives introduced to encourage behavioral change—a major contributor to the process of learning to be sustainable.

5.5. Method: Toward a More Holistic Approach to Environmental Valuation

Problems do not arrive on the public agenda with well-defined dimensions and clarity with respect to which processes threaten which values. Affected parties, who typically harbor differing values and who view the world from differing perspectives, face both substantive and communicative differences. As was aptly noted by Meadows (and mentioned in section 2.3), problems are often articulated not by analyzing what has gone wrong and which values are threatened but by insisting on a particular, favored solution. Favored solutions, it turns out, are usually laced with ideology and propped up with dogmatisms. A change in methods is clearly called for. The key to the new approach provided here is to propose—and then to offer for testing and experimentation—various heuristics and tools that, while not providing prescriptions for substantive outcomes, are hypothesized to encourage cooperative action even in difficult situations. According to this approach, we can expect better and improving outcomes if we introduce and develop problem-solving processes appropriate to the contexts in which problems are addressed.

Pragmatist epistemology—an epistemology based on communal "experience" and corrigibility before the tribunal of experience—deploys the scientific method as a tool by which communities respond systematically and reasonably to complex and controversial problems. In this section, the epistemology developed in section 3.6 is applied to further develop this new approach.

Once cut loose from the mythical search for final truth in correspondence to reality, processes will be judged by their results in action. Cooperative agreements to act, to experiment, and to learn from experience can be expected to be the first indicators of improvement. Pragmatic epistemology places faith in experience—experiments when possible and careful observation otherwise—though success in managing adaptively through collaboration will be manifest in achieving cooperative actions.[3]

Following Adapt, not Optim, one cannot define correct outcomes in advance; instead, one can be content to act on provisional goals and plausible hypotheses, expecting that the experience of the consequences will provide evidence about the action's efficacy. Since both goals and beliefs about how to achieve them are to be judged by the flow of experience, an emphasis on "objective" science is not appropriate. Communities must deliberate, act, and then pay careful attention to outcomes. Hypotheses are judged by whether they help solve problems. Adaptive managers expect progress to be irregular, but they also expect experience—if carefully noted and compared with controls by all parties—to encourage cooperation and to result in a spiral of cooperation and further trust.

To put the point methodologically, following Dewey, I adopt a broad conception of hypotheses. Just as assertions that a given action will have a predicted effect can be judged by impending experience, assertions of value can be judged by acting in pursuit of a value and then observing the consequences of that pursuit. To take a mundane example from my home city, Atlanta, consider the public discourse regarding its "traffic problem." This being a paradigm case of a wicked problem, a comprehensive understanding or final solution to this mess is impossible. One could, as do many current analyses, leave unquestioned the widely expressed value of unlimited mobility in single-occupant vehicles, and could examine the possible means to achieve that goal. This strategy, dominant for decades in Atlanta, has resulted in huge investments in highways and automobile-oriented infrastructure, yet it has failed to prevent gridlock or reduce average commuting time. This trend—not exclusive to that city—has been noticed by traffic experts and has led them to conclude that more roads (more lanes, infrastructure to allow greater speed, and so forth) will lead to more traffic. When new roads are built, more people drive, more people move farther from their work, and gains from road-building fail to support the assumed value of unlimited mobility in single-occupant vehicles.

The pragmatist or adaptive manager, who sees the problem more holistically, refuses to give a free pass to the value or preference creating the complex problem. Decades of experience in pursuing that goal has not led to satisfactory outcomes. At some point, according to Dewey and to adaptive managers, the value or preference itself is refuted by experience. Dewey's point is that to say something is *desired* is to hypothesize that it is *desirable*. Experience with consequences can force us to question this hypothesis and consider pursuing alternative values, such as greater mobility through the cooperative creation of a better public transportation system. Examples such as this illustrate the complexity of wicked problems: there is no a priori means by which one can determine whether to continue building roads, thereby trying to get ahead of the wicked problem. At some point of frustration, it makes sense to criticize the value commitment itself, which becomes increasingly difficult to defend as more and more resources devoted to roadways fail to achieve satisfaction. Similarly, Leopold learned that, while an unnaturally large deer herd was desired, it was not desirable once achieved because seeing deer starve proved it to be self-defeating.

Adaptive management, employed in scientific management of the environment, thrives on the idea of a unified criterion for evaluating hypotheses, whether descriptive or prescriptive. Whether formulated as hypotheses about what is valued or what can be accomplished, all disagreements are to be adjudicated by experience. Dewey's conception of experience was not as individual experience—not gathered as free-floating images in a mental sensorium, but rather collective experience. The advantages gained by creating a broad base of participation in processes require that the process respect the timeworn knowledge stored in the experience gleaned from fishers, farmers, foresters, hunters, and indigenous communities, among many examples.

If, however, such a process also includes sincere deliberation about what are the favored opportunities and the highest aspiration of the human community that has evolved with a natural history—its place—then true social learning can occur. In effective public procedures there must be a place for the dynamic evaluation and reevaluation of the community's sense of its multigenerational public interest.

CHAPTER 6

Heuristic Proceduralism: A General Method

In earlier chapters, it was argued that environmental ethics and environmental economics, as the fields are understood today, and given the questions they address, do not provide appropriate evaluative tools for adaptive management processes. The criticism there was based on practical criteria: as long as the two primary disciplines devoted to the study of environmental values remain conceptually and ontologically at cross purposes, it will be impossible for adaptive managers to use them as starting points to work toward shared methods of evaluating change. Although mired in a cross-disciplinary conflict that currently polarizes and confuses discussions of environmental goals, these fields do have an important role in any full-fledged adaptive management processes. Their failure to advance environmental goals occurs because they are assigned an inappropriate role in the public process. As it stands, the central value concepts of these competing fields are treated as assertions regarding the general nature of environmental values. For example, Economists argue that all value can be represented as willingness-to-pay (WTP) for a given state of the environment, while many environmental ethicists dismiss all economic values as instrumental only, arguing that truly environmental values must be intrinsic to nature itself. These theories are characterized in table 3.1, in chapter 3. Understanding these fields as providing monistic and comprehensive accounts of all environmental value—even as they assert contradictory assumptions about environmental values—creates conflict and polarization. Once one is locked into a belief that a given theory of environmental value is the universally correct characterization of all environmental value, and once one enters the public debate intending to defeat all alternative views, questions as to who has the correct theory of value dominate the discourse. One cannot, on the monistic assumption, begin to provide an overall assessment of the value of an environmental change until one has defeated advocates of all competing theories.

In this chapter, I propose an alternative way of understanding the role of these two competing fields, and I then generalize from this argument to discuss, more generally, the proper role of technical discourses within the public discourse on environmental policy. Identifying these proper roles requires an examination of the multiple, overlapping discourses engaged in debating environmental impacts and environmental policies.

6.1. Dynamic Evaluation

As a first step in unraveling this confusing situation, let us, following Amartya Sen, say that the discourse of environmental values and policy, while existing in a broad form of public discourse based in ordinary language (studded, at times, with technical terms), currently includes at least two "spaces" (or "subdiscourses"; Sen, 1999). The two, or possibly more, intellectual spaces within the broader discourse are based on incompatible assumptions—including, for example, that all environmental values can be expressed in terms of WTP for certain commodities or that "intrinsic" values in nature trump mere instrumental values in important situations. Attempts to amalgamate the reasoning from these distinct spaces necessarily encounter directly contradictory statements about what is environmentally valuable. Obviously, this situation of conflicting assumptions is unproductive, particularly given that adaptive managers believe that goals and objectives must be constantly discussed and modified within a cooperative discourse.

Given these clarifications, can one find a useful role for both economics and ethics in public environmental discourse? My answer is that their "place" is not to provide a universal, top-down value that determines the shape of all evaluative discourse; rather, it is to contribute reasons and arguments within the discourse that may prove convincing to participants who are considering and reconsidering a variety of values. I understand these conflicting theories as providing reasons based in ethics, reasons that have appeared to important philosophers to be persuasive, in public discourse. But here there is no assumption that, to be persuasive, a moral rule or principle must hold generally. Instead, if one is a pluralist, one expects differing rules to be persuasive in differing particular situations (Stone, 1987). Now the problem is not to fit all actions to a single rule, but rather to find multiple rules relevant, and sometimes persuasive, in particular situations.

We thus expect that advocates of these approaches will use them to provide arguments, among others, that may persuade other participants in particular situations. For example, in a given situation, we might hear a mixture of arguments that a problem being faced is like some other situation, whether actual or hypothetical, with arguments that—given those similarities—a specific moral theory should be persuasive. I proceed to show how theory-based methods can be embedded within the adaptive management process, rather than functioning as offering constraints on discourse based on monistic theories. As such, methods of economists and environmental ethicists function, within an open-ended discourse, to provide reasons to favor some actions or policies over others in specific situations. Their role should not be to provide a universal, top-down value that determines the shape of evaluative discourse; instead, it is to contribute reasons and arguments within the discourse that may prove convincing to participants who are considering and reconsidering competing values.

By seeing these arguments as effective within defined and limited intellectual spaces, spaces bounded loosely by reference to particular metaphors, one can forestall direct contradictions and incorporate arguments based on different foundational assumptions. Understanding how theories and the assumptions they embody can create multiple intellectual spaces reveals a new role for theoretical argumentation. So these theories will not be given the role of directly deciding what is rational policy substantively; their role will rather be to encourage deeper thinking about values among citizens and to support a reflective mood of deliberation within a community.

Before reconstituting the two general theories as contributing to an ongoing, dynamic conversation regarding what environmental values— especially those affecting the future—should be, I begin by noting that both theories set out to place a value on *entities*. For economists, value in nature can be represented as *commodities* derived from nature that might be purchased in markets. While they disagree among themselves regarding *which* nonhuman entities have intrinsic value, environmental ethicists identify entities such as living organisms, ecosystems, or species, and they attribute special, intrinsic value to them (Norton, 2005). This shared assumption plays a crucial role in blocking the way forward to a more comprehensive and integrated discourse about human values.

Somewhat irreverently, I refer to the process of breaking a complex, dynamic system into elements that can be listed and attributed value, as "chunking." Descriptions of complex, dynamic systems—especially ecological ones—are made up of separable objects only by abstraction. Seen from the viewpoint of thermodynamics and systems thinking, objects within complex systems are merely markers applied to a process that supports an "attractor" for some period. As one looks at an evolving system over longer and longer periods, more and more apparently static objects erode and disappear over deeper and deeper time. In the quotidian world, one seldom insists on this dynamic aspect, because often the fading of an attractor and the erosion of an associated "entity" will be so slow, when compared to one's sense of time duration, that its temporariness will not matter. I do not, on learning that the boulder in my backyard is temporary—certain to weather away in a few million years—decide thereby not to make it the centerpiece of my gardening efforts. From the viewpoint of nature as a complex, many-layered, dynamic system, then, objects, ranging from boulders to populations to species, are temporary agglomerations occurring within self-organizing systems that support a membrane, creating what can for convenience be called an entity. Objects on this view are supported by processes unfolding in time but are likely to eventually pass into another form, participating in new processes and unlikely to be reconstituted in any of its prior forms.

Against this chaotic backdrop, regular humans make choices and economists and environmental ethicists propose theories, and somewhere in that process, the "blooming buzzing confusion" William James described is miraculously carved up into elements. In economics, the application of market principles cuts nature into commodities, chunks that are or could be traded in markets. "Natural resources" are thus measured in tons or cubic yards just waiting to be assigned a WTP by the choices of actual or hypothetical buyers in actual or hypothetical markets.

Environmental ethicists of the nonanthropocentric persuasion likewise attribute values of nature to objects. A reading of the literature by anti-anthropocentrists reveals a plethora of candidate objects that might be thought to have "intrinsic" value. A preliminary list includes individual organisms, species, ecosystems, and perhaps populations. Further, there is no shortage of earnest arguments that one or the other of these elements in the list is primary or that some are less important. [1]

So, it seems that both economists and intrinsic value theorists have developed evaluative methods that can be applied only after the natural world has been divided into discrete chunks. To evaluate change on such an approach, one must assign values to the discrete entities that exist both prior to and after the change, aggregate these values together, and compare the resulting values in the before-and-after accounting exercises.

The attractions of the quotidian, object-based ontology are obvious for many purposes. The pace and scale of change often does not matter to us, since our interests focus on aspects of a dynamic system that are stable for time frames within human ken. If environmental management is to be intergenerational, however, as posited by sustainability advocates, then it is necessary to adopt the dynamic perspective and realize that, in evolutionary time, which sets the context for ecological time, we encounter "that incredible sweep of millennia which underlies and conditions the daily affairs of birds and men" (Leopold, 1949, p. 96).

Economism and intrinsic value theory (IVT), as currently advocated, embody unidimensionally object-oriented ontologies. They are not good analytic frameworks for adaptive managers for whom nature is best seen as a web of dynamic processes woven into a complexity of multiple layers. This understanding is essential if we are to do justice to the central theme of this book: constant change. Object-based analogies may be useful when attention is paid to the short term and the local, but when sustainability of nature and natural systems is in question, a more dynamic and open-ended, systems-based understanding is necessary.

One might explain the danger involved here by referring to the "Humpty-Dumpty effect." Once that marvelous egg, Humpty-Dumpty, had fallen, "all the king's horses and all the king's men could not put Humpty-Dumpty back together again." And once the marvelous, complex, systems of nature are broken up, there is a value lost that cannot be regained by aggregating them back together again. The only way to avoid the Humpty-Dumpty effect is not to "chunk" nature in the first place.

It is possible to shift to an approach to evaluating environmental change that, being sensitive to multilayered temporal dynamics, evaluates changes *in process*, rather than assigning value to parts or aspects of systems. This shift makes it possible to explore a process-based approach to evaluating environmental change. This approach can be referred to as a "Practical/Procedural/Evaluation/Method," or PPEM, and it is summarized in box 6.1. The PPEM would include, in some form, the steps or elements listed in box 6.1.

Box 6.1. The Practical/Procedural/Evaluation Method (PPEM)

1. Describe several possible paths into the future, and associate each with a set of policies or actions that would make that path more likely, using scenario development and other techniques to encourage speculation regarding possible pathways to the future.

2. Associate those paths with characteristics that can, in turn, be treated as measurable, preferably synoptic, indicators. Indicators are empirical measures of a process that is associated with the production of a social value.
3. Generate public discussion of the comparative importance of the various indicators.
4. Guide discussion toward actions that favorably affect those indicators that score highest on the importance ranking, treating these as elements in a multicriteria decision model.
5. Continue discussion, iteratively, seeking policies that can be expected to improve at least one important indicator, with cascading positive—or at least neutral—impacts on other indicators.
6. Study outcomes scientifically, employing the method rigorously, and evaluate actual outcomes.
7. Seek opportunities to find common ground in actions, and build coalitions to experiment and explore options. Cooperative action even on small issues can create trust and expand opportunities for further progress. If one denies the possibility of chunking nature, one is, by that simple act, apparently entering new territory in the incipient field of evaluation studies. Again, it seems a void is encountered. If one does not chunk nature and apply one of the entity-oriented theories of environmental value, how can environmental change be evaluated?

The void should be a lot less scary, now that the process of decision making has been supplemented with some concepts and heuristics. The situation, on the contrary, is ripe for a breakthrough. Once it is acknowledged that chunking is futile, evaluators must turn to another unit of analysis: in this case, I recommend that practitioners in the new field of evaluation studies direct their attention toward development paths as the key unit of analysis for evaluating environmental change. A development path, which describes a possible path into the future for a community, can be thought of as a kind of scenario that can be expected to unfold if certain policies and strategies are exercised. From any point in time, a placed community will face a number of possible ways forward, and by comparing these and choosing, it might over time be said to steer adaptively on a particular path toward sustainability.

This step is relatively simple. Baffled by the polarization between economists and environmentalists over which objects can have value, and recognizing the need for community involvement and more holistic thinking, I suggest that communities shift the question, stop trying to associate environmental value with fragments of natural systems, and begin comparing possible development paths more holistically. Once so focused, they can begin to identify some attractive paths forward and to associate them with various possible policies and interventions that might lead to the favored paths. This focus will, in turn, make clear that there are many dynamics and many choices open, and discussants will no doubt tell many stories that explore alternative scenarios. The point is to shift thinking from placing value on elements of the system forming a "situation" to exploring and evaluating various scenarios that could unfold if myriad actions and policies were undertaken in that situation.

So, if one asks, "How can chunking be ended?" the answer is absurdly simple: Reject the assumptions and the perspective that sees nature as a set of commodities or as independently valuable entities. Stop thinking of "value" as a noun and think of it as a verb. Reject, further, the even deeper, monistic assumption that one conceptual framework can or must fully capture all varieties of natural value. Now, with a mind cleared of rubbish, ask, "How can a community of diverse individuals and groups express their values in a more holistic way?" This question provokes the following, plausible possibility: individuals and groups will express their values in the form of comparisons and rankings of various possible development paths that could lead them into the future. In the remainder of this chapter, we will explore how PPEM can be developed into an important element in a broader, adaptive management process.

6.2. Discourses, Spaces, and the Place of Technical Languages

First, before putting moral and economic theories of environmental value in their proper place within the adaptive management discourse, this section provides a conceptual map of the discourse regarding environmental values and the intellectual space in which human values are translated into recommendations for action. The current discourse regarding environmental values and policy is a confusing hybrid of ordinary language laced with technical terms developed in a variety of specialized and academic disciplines. As a result, one encounters (1) discussions mostly in ordinary language, in which speakers and listeners may understand each other, but often imprecisely, as participants operate with different definitions of key terms, which are given no technical definition (discourse type 1); (2) disciplinary discourses in which ordinary language, while unavoidable, is in important contexts replaced with the technical language of a discipline that, in turn, invokes the assumptions shared within that discipline (discourse type 2); and (3) hybrid discussions that, while apparently carried on in ordinary language, include technical terminology drawn from multiple disciplines, such as economics and environmental ethics (discourse type 3). Discourse types 2 and 3 differ in that the former exemplifies a discussion in which technical terminology from a single discipline shapes ordinary language into a monistic system such as economics or ethics. This discourse type can be useful, but I have shown that these monistic systems can never describe the true situation in all its complexities. Discourse type 3, on the other hand, incorporates into ordinary discourse technical definitions and assumptions from multiple disciplines. Since it has been shown that these assumptions can conflict, discourse type 3 is pluralistic but subject to confusion because of conflicting assumptions built into different elements of the discourse.

To sort out this confusing situation, we have borrowed the idea of "spaces" within discourse from Amartya Sen (1999), noting that there are multiple spaces that permit intelligible conversation about environmental values among groups of discussants who accept, either permanently or temporarily, the assumptions that give intellectual shape to the discourse in that limited space, with the space delineated by particular problems and concerns. Given the imprecision of unreconstructed ordinary language, progress toward better understanding and better decisions encourages movement either into discourse type 2—adopting monism by characterizing and addressing problems assuming a single discourse shaped by a particular discipline—or into discourse type 3, which creates a space in which technical languages from multiple fields are mixed together, raising the possibility of conflicts among assumptions imported with the technical languages used. Discourse type 3 seems more complete than discourse type 2, but it is problematic because importing a mixture of disciplinary jargons from multiple technical fields can result in confusion and contradictory assumptions.

This conundrum—of seeming to have to choose between imprecise ordinary language (discourse type 1), limiting one's technical imports of more precise language to one or to compatible disciplines (discourse type 2), or embracing technical assumptions and definitions from multiple fields (discourse type 3), at the risk of embracing contradictory assumptions derived from different technical languages—can be managed by appealing to Sen's concept of limited spaces within general discourse. By separating discourse into multiple spaces, one can endorse the development and use of multiple type-2 discourses but avoid the contradictions threatened if we embrace discourse type 3, which unsystematically imports incompatible assumptions from multiple disciplines.

Spaces are constituted by the concerns and the topics that motivate them. For example, if one is working on toxic pollution problems, it no doubt makes sense to introduce concepts from the field of risk assessment and management (creating a space in which individual risks come to the fore), whereas a different set of concepts such as HT and resilience can be introduced if the problem and concern centers on the declining integrity of ecological systems. The spaces, in other words, provide a site where one can become technical about factors affecting efforts to deal with limited clusters of problems. This approach, based on Sen's idea of spaces, allows us to embrace the existence of plural sets of problems that may require different tools borrowed from different disciplines. If, however, these multiple type-2 discourses are applied within a limited space, they may each be useful in some spaces, while not requiring that they apply across the board to all spaces.

Thus, if one wants to progress past the imprecision of ordinary language and create a more sophisticated discourse, developing multiple type-2 discourses applicable in, but limited to, different topic spaces is the most appropriate strategy for adaptive managers. Progress is made by developing discourses borrowing from the special disciplines for important ideas and concepts but avoiding mixtures of concepts that will result in contradictory assumptions by applying these disparate discourses of type 2 in the same spaces. For example, in one space all value discussed is anthropocentric value, and terms and concepts appropriate to that discussion are used in that space, without any commitment to make concepts used in that space consistent with the concepts and ideas in other spaces (such as discussions of animal rights). Spaces, unlike disciplines, offer specialized vocabularies for specialized purposes, and the concepts used in one space need not be consistent with all the disciplinary assumptions used to address the full range of environmental problems. This framework of discourses and spaces squares with and operationalizes pluralism: it embraces multiple technical frameworks but does not attempt to build these into comprehensive accounts of all environmental problems or concerns.

I have referred to "technical languages" that are imported into environmental policy discourse and have cited mainly environmental economics and environmental ethics, though the importations are not limited to these two fields. Distinctions and concepts find their way into the discourse from several technical disciplines, such as risk assessment and systems ecology, in addition to environmental economics and environmental ethics. While I cannot discuss in detail all these technical augmentations of ordinary speech, the general idea can be introduced by surveying the strengths and weaknesses of the space in environmental discourse dominated by the market metaphor. In this space, environmental value is assumed to be measured as WTP, while benefits are interpreted as goods enjoyed by individuals who are assumed to choose and purchase commodities solely on the basis of their perceived self-interest.

Economists' concepts are designed to aggregate individual welfare (and thus to influence policy by comparing costs and benefits of open choices) and thereby to describe human behavior and preferences as they presently stand. For these models to make sense, theoretical concepts, such as temporal discounting, are assumed so as to regularize all costs and benefits to current dollars and to compare monetized valuations across time. As economic models are imported into this space, discussants become wedded to "equilibrium" models that measure values only at the margin. It is further assumed in this intellectual space, in particular, that preferences of individuals are fixed over time.

In this space, where the task is to understand present preferences, there is no possibility of a conceptually coherent discussion of how individuals and groups might change their preferences or adjust their objectives. Nor is there room for models that move, dynamically, from one functional state to another. So, attempting to understand dynamically changing environmental values cannot be accomplished in this space, where the task is to describe current preferences. It cannot explore the possibility that, over time, a more "educated" and more "appropriate" set of preferences might evolve. Further, since these models assume that each individual acts out of self-interest, the intellectual space defined by these assumptions may introduce bias against altruistic or community-oriented values, which will have to be explored in a different space with a different set of tasks.

Adaptive managers, who expect that collaborators will have to learn their way toward sustainable living and that this learning will include vetting and perhaps reforming preferences along the way, are not likely to embrace economic modeling as an adequate discourse by which to understand environmental valuation in a dynamic manner. The space within environmental discourse that accepts and incorporates assumptions from mainstream welfare economics, then, studies and aggregates current preferences, yet leaves no room for discussions of how, why, and whether preferences of individuals should be modified.

Interestingly, environmental ethicists who have adopted the theory that some elements of nature have intrinsic value operate in a space in which they find it necessary to criticize and replace many of the population's current, anthropocentric preferences with less human-oriented ones. This space in the discussion of environmental policy excludes many concepts and considerations that economists find necessary to their adopted task, and will introduce alternative assumptions and terms that are more appropriate to a task that includes understanding changes in values expressed. The two spaces inhabited by advocates of these different spaces essentially exclude the other from intelligible discussion, though this is tolerable if one assigns them somewhat different tasks and roles.

The intellectual space constituted by economic assumptions appears inadequate to environmental ethicists, given their goals: they hope to change present preferences, a very different task than that undertaken by economists and ethicists who have found the economists' space unsuitable to their tasks. But ethicists explain this inadequacy in two rather different ways. First, many environmental ethicists actively reject anthropocentrism and refuse to even enter the space formed by economic assumptions. This group has a large following among philosophers and activists, and operates within an important space constituted by assumptions contrary to economists' assumptions. Among this group, there is no ground for compromise with mainstream economists, because merging the space of economics and environmental ethics requires embracing contradictory assumptions.

There is, however, also a broader ethical critique of the market model that need not invoke nonanthropocentrism. Other ethicists reject the market model on the alternative ground that the assumption of preexisting and unchanging preferences is the most serious flaw in the economists' discourse. Leaving aside the advocacy of nonanthropocentrism, it can be argued that the economic space is inadequate to include *any* moral program that engages in moral critique of the values and preferences of dominant groups in society. Thus, any discourse that freezes preferences by importing assumptions of economic valuation is inadequate for any ethicist who believes, on whatever grounds, that current preferences must change. On this criticism, ethicists—whether they are anthropocentric or nonanthropocentric—might appeal to a wide range of values, such as obligations of fairness to future generations, concern that national policies harm the poor, or duties to other species. Moralists advocate changes in the moral beliefs and actions of current populations. Any of these concerns require that human agents change their preferences, so a commitment to dynamism regarding values can motivate the occupation of a space different from the preference-measuring that engages economists.

This space, in which values are considered dynamic and open to change, need not lock into the assumptions and concepts of anthropocentrism or nonanthropocentrism. Rather, in this space, a coalition can be formed of those who, whether anthropocentrists or nonanthropocentrists, endorse a diverse and morally rich discourse that engages in critiques and advocates changes in current values. In this intellectual space, the search for a multi-generational public interest must be understood as going beyond counting the static, fixed preferences of today's population. Such a search seeks to encourage values that emerge over time as the public engages in deliberation and social learning. In this space, the shared task is to understand preference change, a task that can be undertaken while leaving open for further learning any conclusions as to what would be the "correct" value commitments. This space will be extremely important as adaptive managers broaden their process to include not only a pursuit of scientific insights but also social learning regarding objectives and values.

One can now see that, by paying attention to various intellectual spaces within broader environmental discourse, and associating them with different tasks, one can say that the space occupied by economic discourse may be the appropriate space when the goal is to estimate what the public's current preferences are—a relevant type of information. Findings and empirical work about current preferences may thus provide important data to consider. If, however, the question is a long-term one about which broader values should be sustained, it will be necessary to move out of the economic, fixed-preference space and into a dynamic space in which it is assumed that environmental values can, and in many cases must, change. No single intellectual space can address all questions regarding environmental policy or sustainability; yet many different spaces are available to encounter, reexamine, and revise current values, provided we regard them as open to change within a system evolving toward sustainability.

So, the "place" of technical and professional considerations and language within an adaptive management process is not to provide a reduction of complex questions to simple ones, all with the same form. Their place, instead, is to serve as tools deployed within an ongoing process of adaptive management and social learning, tools that can shape useful spaces for exploring a wide range of specific questions both descriptive and evaluative. Since the goal here is not to identify a single, comprehensive approach that will work in all situations, but rather to learn in response to many different challenges, there is no reason to attempt to enforce a common language that will be appropriate to every intellectual task and associated space. Here, favoring discourse type 2 applied in a limited space with special assumptions imported from economics allows the importation of insights from scientific disciplines, while avoiding direct contradictions. Ideas and concepts from the special disciplines thus find their place both as catalysts for new ways of thinking and as aids to the emergence of an environmental consciousness complex enough to engage the problem of finding a sustainable lifestyle.

A space in which values and preferences are dynamic is essential for an environmental policy discourse that is rich enough to contribute to the broader goal of adaptive management—that is, learning the way toward sustainability. That process must be undertaken in a dynamic intellectual space in which many values can be expressed and the stringent methodological constraints applied in the economic space will not be enforced. Recognizing different spaces for different tasks frees adaptive managers from the requirement of monistic thinking—that there must be one correct theory and one associated correct language for discussing environmental goals and values. In the development of tools and spaces in which to use those tools, one should let many flowers bloom by limiting the scope of application of introduced technical terms so as to avoid the contradictions that occur if we try to use these technical terms comprehensively.

Heuristic 8

The dynamic evaluation heuristic. Never assume that values are fixed. Adaptive management provides opportunities to question and re-form values in process through social learning. Dialogue and deliberation should be inclusive, but shaping of values should aim at articulating and pursuing the emergent public interest.

6.3. The Capabilities Approach

Identifying and separating these intellectual spaces makes explicit the linguistic resources available for a truly successful process of rethinking and re-forming public values and action regarding the environment. Trying to include economic thinking and ethical thinking in the same space creates conceptual chaos because the two vocabularies carry with them assumptions incompatible with assumptions accepted by others. By organizing discourse into spaces with different emphases and focused on different tasks, technical language and concepts can contribute to particular tasks, while the various spaces are sufficiently isolated from other spaces to avoid contradictions and discipline-based conflicts.

This positing of multiple spaces may at first seem chaotic, but it can only help to be more explicit in addressing value and preference change. What is needed is an intellectual space in which the language is diverse enough to accommodate pluralism of values but that also embodies certain organizing principles and concepts that will help people think about what they want from their environments. Fortunately, there is an emerging approach— associated with the work of Amartya Sen and sometimes called the "capabilities approach"—that provides a way of connecting people and their needs with the opportunities and limits of the environment they live within. The capabilities approach can readily be used, and has been used, in development ethics; consequently, incorporating it here may help expand thinking from "sustainability" to "sustainable development."

The capabilities approach provides a space in which, unlike in the economic space, value emerges in situations of choice rather than from fixed preferences. The two main tenets of the capability approach are (1) that the primary moral value is the freedom to achieve well-being (as understood by the subject) and (2) that the freedom to achieve well-being is best understood as a matter of capabilities. Moral well-being thus depends on the real opportunities that individuals have to do and to be what accords with their values (Robeyns, 2011).

Sen sees the capabilities approach as an alternative to forms of utilitarianism that restrict "the judgments of states of affairs to the utilities in the respective states (paying no direct attention to such things as the fulfillment or violations of rights, duties, and so on)" (Sen, 1999, p. 62). He thus rejects attempts to understand human well-being purely in terms of subjective feelings and preferences, insisting instead that true freedom requires individuals to have actual abilities to achieve well-being within situations in which they pursue their goals. Evaluation within this broad approach thus takes into account not only subjective welfare satisfaction but also the values that rational beings set for themselves and individuals' abilities to succeed in various situations. The capabilities approach directs attention to information that is essential to judge how well a person's life has gone, and it also creates a space in which comparisons can be made across people and, for a single person, across time.

In this intellectual space, one speaks of "functionings" as morally neutral achievements: so gaining an education, or being part of a helpful social network, are functionings, but so are contracting a dangerous disease or becoming a gang member. Capabilities are an individual's real opportunities to achieve various functions, yet functions must be evaluated with reference to the individual's personal values and life goals. "Functionings" and "capabilities" can thus be proposed as metrics for comparisons of well-being, with functionings being a measure of achieved well-being and capabilities being a measure of the freedom to pursue that well-being.

Still, it is important to recognize that the capabilities and functionings approach, while helpful in identifying the goods and services needed for true well-being, also emphasizes differential abilities among persons to make use of those goods and services by converting them into functions. A person without elementary education will not be helped by the availability of college scholarships, for example. Sen refers to these aspects of the achievement of well-being as "conversion factors," which is the degree to which a person has the actual ability to deploy a resource to obtain a "functioning."

Three types of conversion factors are often noted (Robeyns, 2011, sec. 2.4):

1. *Personal conversion factors* are characteristics of individuals that make the turning of capabilities into functions either easier or more difficult. Offering a college scholarship to an illiterate person has no effect because the person lacks the relevant conversion factor. But being seven-feet tall may be a conversion factor that will be effective in garnering a college basketball scholarship. This concept is related to another important term in development ethics, "capacity." Capacity building is development that enhances agents' personal conversion factors.

2. *Social conversion factors* are conditions of the society in which one encounters a capability. Being low on social hierarchies such as class or caste can block certain individuals from turning capabilities into functionings, just as widespread liberal social policies can help many people to survive and transcend poverty. The appropriate response to such blockages is often to engage the entire society in "capacity building."

3. *Environmental conversion factors* are factors in the built, or artificial, environment *and also in the natural physical and ecological environment* that can make the conversion of a capability into a functioning either possible or impossible. Being in a resource-poor environment can make conversion of important capabilities into functionings more difficult, just as living in a resource-rich environment makes conversion easier.

Sen's reasoning rests on a separation of two senses of personal freedom: process freedom and opportunity freedom.

> Freedom can be valued for the substantive opportunity it gives to the pursuit of our objectives and goals. In assessing opportunities, attention has to be paid to the actual ability of a person to achieve those things that she has reason to value. In this specific context the focus is not directly on what the processes involved happen to be, but on what the real opportunities of achievement are for the persons involved. This "opportunity aspect" of freedom can be contrasted with another perspective that focuses in particular on the freedom involved in the process itself (for example, whether the person was free to choose, whether others intruded or obstructed, and so on). (Sen, 2002, p. 10)

Both forms of freedom are important, though opportunity freedom is of direct importance here because it connects values to the environment by including—in conversion factor 3, above—the natural resource and natural systems aspects of environmental conversion factors that determine whether individuals can achieve valued functionings. The capabilities approach thus accounts for the importance of natural resources and natural processes in the achievement of functions essential to well-being. This approach may provide a starting point for a new approach both to assigning value to environmental features and to effecting changes in them. In this approach, changes will be considered positive (other things being equal) if they support human capabilities and negative if they do not.

This approach to evaluation of environments and changes in environments fits nicely with the arguments of chapter 4, where it was argued that an ecological system or a human habitat is encountered as a mix of opportunities and constraints. Future sustainability of a strong sort must include an understanding of ecological and physical features of the natural systems that future people will populate. Strong sustainability can thus be described within a capabilities approach that recognizes the importance to both the present and the future of protecting opportunity-producing systems, such as forests and fisheries. Environmental and ecological concern for the future can thus build on the capabilities approach, especially a version of that approach that emphasizes the importance of "stuff"—the physical features of an environment that support opportunities. So we can state our sustainability goals in terms of capabilities: earlier generations have an obligation to protect those physical features of a natural system that are important in transforming capabilities into functionings, and functionings into opportunities, as sketched in the schematic definition of sustainability.

By introducing the concept of capabilities as freedoms conceived as real opportunities, Sen is referring "to the *presence* of valuable options or alternatives, in the sense of opportunities that do not exist only formally or legally but are also effectively available to the agent" (Robeyns, 2011, sec. 2.7). Opportunities reflect human values, though they also depend on states of social, economic, and ecological systems. Sen's capability approach, then, applies nicely to the hierarchical models used to define unsustainability in section 4.3, as damage to environments that diminish opportunity freedom. The capabilities approach thus provides a moral space in which goals of strong sustainability can be deliberated.[2]

One reason to describe the capabilities discussions as an "approach" rather than, perhaps, as a "theory" is that multiple theories have been developed within the capabilities approach. One such difference can be found in a comparison of work in the area by Amartya Sen and Martha Nussbaum. Nussbaum, on the one hand, argues that there are certain key capabilities that, if lacking, make the living of a fully human life impossible (Nussbaum, 2006). She lists ten capabilities, including, for example, health, bodily integrity, and the ability to control one's environment. Notice that this way of thinking shares several features with Optim: the idea is to specify a good life in advance, by analytic techniques, which then puts emphasis on finding the means to fulfill that good life. While Nussbaum's list of ten certainly suggests a more robust goal for social planning than, say, to optimize economic efficiency, the list also evokes Optim's efforts to set goals in advance of action.

Sen, on the other hand, operates more in the spirit of our character Adapt. Sen does not believe there is a canonical list that applies everywhere, and he thinks the most appropriate list may vary, given the specific problems addressed. He does not think it necessary to seek an a priori list of capabilities based on analysis or appeal to first principles. He expects, on the contrary, that judgments regarding capabilities will emerge democratically from a participatory process perhaps similar to the one developed in this book. For this reason, I follow the approach developed by Sen.

While it is no doubt premature to declare that capabilities approach will provide all the normative foundations of adaptive management, it does define a moral space in which to discuss, deliberate, and engage in social learning. This approach can be recommended as providing the appropriate, dynamic context in which policies are evaluated. It represents a step toward creating a new, open-ended intellectual space where environmental values can be discussed in a dynamic context.

The capabilities approach, however, is not intended to be merely one tool among other tools. It is, rather, a way of thinking through problems in which human values (broadly construed) are discussed against a backdrop of assessing real opportunities and trends in their loss or protection. The capabilities approach, featuring an open-ended search for sustainable development, provides a space in which to develop tools for analyzing real opportunities against the backdrop of ecological realities.

Heuristic 9

The capabilities heuristic. Evaluate environmental change from the viewpoint of broadening capabilities of present *and* future people. Paying attention to capabilities emphasizes freedom of choice over many opportunities to achieve well-being, and it highlights ways in which physical processes support the multigenerational public interest.

6.4. The Role of Specialized Disciplines in Adaptive Management Processes

The goal of adaptive management is, whenever possible, to turn vague or ideological differences of opinion among participants into hypotheses that can be tested. This goal implies that the empirical sciences must play a key role in any adaptive management process. The democratic and participatory aspects of adaptive management, however, place important constraints on the ways in which scientific hypotheses and scientific findings are communicated to process participants. If a scientific finding is put forward as implying an important change in policy, democracy's demands imply that those findings should be intelligible to all members of the process, which places a heavy burden on scientists as interpreters of empirical findings.

6.4.1. Properly Placing Economics and Environmental Ethics

In this subsection, I will now use the idea of intellectual spaces to identify the proper place for economics and environmental ethics in deliberative and adaptive processes. It has been noted that economists have a single measure of value—WTP, which brings with it the assumptions that preferences are fixed and unchanging. Strict adherence to the definitions and assumptions of mainstream economics creates a space in which any discussion of how and why values might change across time has no place. If one believes that individuals in contemporary society need to modify their values and preferences, their challenge must be developed in a different space than the disciplinary space defined by economists. For that task, it is necessary to step outside the static space defined by economists into a space that allows dynamic alteration of both assumptions and values.

Environmental discourse, in its current state, is an unwieldy public discourse that includes conceptually inconsistent concepts and assumptions. Disciplinary economists have staked out a space in which they can make progress in *describing* preferences, using technically defined concepts to estimate and aggregate what members of a population prefer. Within their conceptual space, economists can provide some measure of people's actual values, at the present time and also insofar as they can be observed in human behavior. In their space, economists engage in static evaluation of environmental changes.

Knowing the public's current preferences can, on many occasions, be useful to adaptive managers. For example, if they must distribute limited resources in support of recreation, or when deciding how to invest in a scientific pilot project, it can be useful to know what the people prefer. Learning which sites would garner more users and more public involvement could help determine which site to restore first. Adaptive managers, however, cannot limit their discussion of values to a static decision space. Because those managers do not yet know how to specify the goals of sustainability in advance, and because they expect discussions of what is valuable in particular contexts to lead to lively debates, they need an intellectual space in which deliberation about values and the effects of present preferences can lead to experiments and learning. Progress toward sustainability can only occur in a dynamic discourse where participants deliberate and either reaffirm or modify their values and the objectives pursued. Thus, while adaptive managers can use information about present preferences in many decision situations, they must also engage within a more dynamic space in which values broader than economic ones are debated and reconsidered.

So far, then, at least two intellectual spaces have been found appropriate to addressing quite different questions. Environmental economists have adopted the task of *estimating* what values—and their strength—actually exist within a population. Environmental Ethicists, on the other hand, have adopted the task of *criticizing* current values and preferences; they have as an explicit objective to cause people to question their values and to perhaps expand them beyond existing values that are too short-term or too narrowly focused on human interests. In this critical space, it must be possible to criticize and attempt to alter current preferences and values, and in this space, a variety of arguments and principles may be introduced. Appeals will be made to the intrinsic value of nonhuman entities, but critics of current preferences and actions might also appeal to obligations to future generations, to principles of virtuous living, and so forth. In this space, the question then is not merely which values people currently have, but which values they *should* have if they are to properly protect nature as well as human well-being on the way to sustainability. This space in public discourse is important in providing individuals with an opportunity to argue for their ethical position openly and by making a case for the actions and strategies they advocate.

In the best of cases, public discourse—especially if it involves experimentation and changes to the structure of decision situations—can encourage incremental learning that can lead to true transformations of perspective and a reorientation of priorities, or what has been called "double-loop learning." As argued by Kai Lee (1994), the ultimate purpose of public discourse is to transform perception of problems and to allow real progress.

Now it will be possible to "place" discussion of values within the larger discourse of environmental science, action, and value. Environmental economists and environmental ethicists provide the conceptual raw materials for an improved and more sophisticated discourse, helping to clarify concepts and to provide clarifying arguments.

The important limitation here is that no specific space has been developed in a way that captures all the value considerations relevant to deliberation and social learning. In part 1, it was learned that there cannot be a single, definitive model for messes and complex problems, so arguments—and the spaces in which they are articulated—will have no arbiter from a single discipline. Instead, they will be adjudicated in the ongoing deliberative process. So, environmental economics and environmental ethics will both be important in environmental policy discourse. Each of them will find an appropriate place in discussions, with economic reasoning being more appropriate when we are trying to accomplish something that is widely desired by a given population, and environmental ethics and its literature being more relevant when it is recognized that some environmental problems, especially long-term ones, can only be addressed when humans come to question and adjust their current values.

Disciplinary tools can create arguments and bring important concerns to bear within a process of deliberation. Spaces provide a patchwork of limited frameworks that evolve to address emerging problems, but they cannot be amalgamated to offer a comprehensive analysis of all problems. Rather, the appropriateness of arguments from particular, limited spaces will be resolved in the court of public opinion. In this "court," there can be no appeals to universal truths. It operates as individual participants discuss and deliberate, often within bounded intellectual spaces, each of which may be judged appropriate, or not, for particular problems and issues. But nothing can be gained by trying to create a superspace that will encompass all problems or all theories, because such a superspace would inevitably embody inconsistent assumptions.

6.4.2. Disciplinary Stew

I can now proceed to the larger question of the role of academic disciplines, including not just economics and environmental ethics, but also all the disciplines that have a branch in one of the "environmental sciences." Recent decades have seen the development of new fields of environmental science, both natural and social: to wit, conservation biology and environmental engineering in the former group; and in the latter, environmental sociology, environmental psychology, and anthropology, as well as economics. This broadening has occurred through the development of new disciplines or the emergence of subdisciplines focused on environmental problems within traditional disciplines. If these sciences are to be effective contributors to collaborative management, then all must contribute to the public dialogue.

At the heart of any such process must be a discussion space, which can be thought of as a public square, or a "table" at which participants can interact. The important thing is that the deliberative space must be open to all interested and affected parties. So, in the view of collaborative adaptive management processes being developed in this book, success depends on the appropriateness of the process to a decision being pondered in a particular context. The reasonableness of outcomes emerging from appropriate processes will depend on the reasonableness of the discourse that informs actions.

In the remainder of this section, I offer the essential structure of a constructivist, process-oriented approach to evaluating environmental change by detailing four aspects of an appropriate discourse guiding public processes forward through a collaborative process: (1) pluralism and a multiscalar vision, (2) ideal speech, (3) disciplinary stew, and (4) embedded science.

By concentrating on these factors, communities entering deliberations about felt problems can design a process appropriate to their needs—one that will encourage dynamic evaluation in a context in which they can seek policies that contribute to opportunity freedom for residents and build community capacity in search of sustainability. Such a system of evaluation would draw on all the sciences contributory to adaptive action and also on relevant humanistic disciplines such as history and philosophy.

6.4.2.1. Pluralism and a Multiscalar Vision

Since (1) wicked problems are not well formulated initially, (2) aspects of these problems can appear on different scales and affect open-ended processes, and (3) many individuals and groups will take different viewpoints on them, modeling and representation of the problem should be scale-sensitive and, if necessary in particular situations, multiscalar in structure.

As noted in the earlier reference to Justice Holmes, democracy is at its best when varied and wide-ranging ideas compete in a public arena; this openness must extend to radical and unorthodox ideas. Having shed the idea of a single "right" answer, I encourage discourse in which citizens express opinions openly, creating diversity and conflicting opinions, and a lively competition ensues. In the words of the pragmatist Bernstein, "Here one begins with the assumption that the other has something to say to us and to contribute to our understanding" (Bernstein, 1997, p. 399).

One might wonder whether a robust pluralism must also be multiscalar. I believe it must, because appropriate problem formulation in the face of wicked problems requires sensitivity to both spatial and temporal aspects of problems. For example, as different groups, organized at different scales, address a problem (such as the placement of a permanent storage site for nuclear waste), different values will be highlighted if one looks at the problem at different scales. The evaluation of the policy choice to site a nuclear waste repository at Yucca Mountain in Nevada looks entirely different, and highlights different values, if one looks at it from an office tower in Washington,

DC, or from a ranch in Nevada. Economists try to reduce every type of value involved in a dispute to present dollars and then try to aggregate preferences across these scales. Pluralists laugh at this fool's errand, recognizing that the values emergent at different scales of a system are often incommensurable. In such cases, it will be necessary to evaluate impacts of changes on multiple scales, seeking a reasonable balance in protecting values across levels rather than seeking a reduction to one comparable level.

6.4.2.2. Ideal Speech

Progress in collaborative, adaptive management is only likely if certain features are exemplified in the public discourse in which management choices are discussed and negotiated. While public speech is never ideal in practice, striving toward an ideal of an open-speech community in which disagreements are aired and discussed by giving reasons, rather than by name-calling or use of power, is essential to successful collaborative adaptive management.

I have defined the "public interest" in terms of an ideal process: the public interest is what would result if the people in a community were to be public spirited, well motivated, intelligent, and inclined to rational resolution of problems. Here, one might also add that the community respects and learns from the advances of the sciences and special disciplines that explore human and natural interactions. If this sounds too good to be true, of course it is. One can only use the ideal hypothetically; yet positing, however hypothetically, an ideal discourse provides the opportunity for comparisons and even ranking of different speech communities by discussing the extent to which they approach the hypothetical ideal.

The phrase "ideal-speech community" is adopted, of course, from Jurgen Habermas, who espouses a communicative ethic and who has catalyzed a discussion of what he calls "post-metaphysical" analysis and understanding of values (Habermas, 1984). I will not go into detail regarding his ethical views here, except to embrace his procedural ideal: in an appropriate, democratic public discourse, all opinions and all opinion-givers deserve respect, in the sense that they are responded to with reasons and evidence, not force or ridicule. One key procedural ideal of broad application in public policy process is equal access of all affected parties to the discourse, understood as a place at the table where deliberation takes place regarding what to do. This ideal, and others, requires the existence of a common language in which their discourse takes place.

It would be futile to require that communities achieve the status of an ideal speech community before embarking on a collaborative process approach.

On the contrary, having an ideal, and realizing how far one is from that ideal, can be the catalyst for discourse, deliberation, and perhaps the emergence of a public behind a cause. While attaining an ideal-speech community is not possible, seeking to create one, and seeking to act cooperatively within one, drives a reflective, self-reflexive, and self-correcting process.

6.4.2.3. Disciplinary Stew

While practitioners of the special disciplines will routinely make use of technical language in their scientific studies, if that work is to have a salutary effect on a discourse-driven, adaptive process, it must be articulated and explained in the ordinary language familiar to participants.

Based on the importance of the commitment to an open public discourse, I am driven to the conclusion that the basic language of adaptive management processes must be ordinary language. There will always be a place for introducing technical detail in a public discourse as explained in section 6.2.2, but the hard-nosed answer, following from the Habermasian ideal, is that such technical detail must be explained and reviewed in ordinary language before it can provide valid reasons to adopt or reject a policy.

The best analogy available—and it may be a bit corny—is to think of the process and its goal as one of creating a wonderful, complex stew, one that will please people from many cultures, but that will also gain accolades from gourmet cooks and famous tasters from all lands. I call this dish "disciplinary stew." It is "disciplinary" because, by analogy, we need some meat and potatoes, some chewy and strong-flavored information, but a stew also must embed those ingredients in a broth. So the specialized and academic disciplines can be thought of as the meat and potatoes floating in a broth. The broth, thinner and less dense, is ordinary language. A stew, while featuring high-protein ingredients, however, is only a success if the flavors and nutrients have given their flavorful essence to the broth—see figure 6.1. By analogy, a public deliberation and decision process requires broad communication but also requires, in some cases, state-of-the-art and highly technical science. Openness and public participation, which is essential to collaborative processes, can only be achieved if scientific inputs are intelligible to the participants.

One more key point must be made here, before leaving the kitchen. It is important that the stew include, in addition to the natural sciences, also social sciences and humanities: ethics, environmental history, environmental sociology, poetry, and art. All these should contribute to an open discourse among citizens with differing levels of education, varying interests, and myriad perspectives. The academic disciplines are all "in the stew

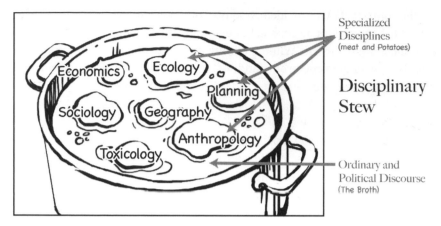

Figure 6.1. Disciplinary stew: Like the meat and potatoes in a stew, insights from the special disciplines lend rich sources of nutrients for public discourse; but these insights must be presented in the "broth"—the thinner and more transparent discourse of ordinary language—if they are to determine environmental policies.

together." That is not news. What is news is that, if adaptive policy is the goal of the academic environmental disciplines, the voices of the special disciplines, while harboring important expertise, have exactly the same status as the other participants and affected parties. If the goal of any of the special disciplines is to contribute to a collaborative, adaptive process, then their success will be measured in their ability to provide useful tools by which communities can explore some aspects of complex problems.

True openness is one of the most difficult of all political goals. Openness—as opposed to the exploitation of one-sided information—is unusual and precious. Building on the commitments explicit and implicit in pluralism and on ideal-speech behaviors, it is possible to place an important constraint on environmental policy discourse and its assumptions. That constraint states that, while cutting-edge science is often of great importance, that importance will only manifest if it is presented, explained, and interpreted in a way that makes it accessible to all participants in a decision-making process.

6.4.2.4. Embeddedness

Ideally, scientific study in adaptive management and deliberation about goals and values should be embedded within a given public policy process; further, these studies should be responsive to the uncertainties and disagreements that stand in the way of cooperative action in the process at hand. Environmentalists, along with many others, embrace "scientific" or "science-based" management. Yet there is great variation in what this means, and even greater variation in the exact role that science is expected to play in addressing complex environmental and social problems.

What it means in the case of adaptive management is that environmental science should be embedded within a public process of planning, acting, and evaluating—that is, an open discourse. Science, in a public, active adaptive management process, will be science in pursuit of a public value, what Funtowicz and Ravetz (1990, 1993) call "mission-oriented science." Of course, there will be many situations in which the scientists working within the management process will identify and make use of studies carried out by academic scientists having curiosity-driven motives; at other times, those managerial scientists will sponsor research by academic scientists, soliciting studies on a topic of importance to management choices. As noted in section 6.4.2.3, such studies may be the meat and potatoes of the process; but if such studies are to be used to justify decisions, then the results must be presented to managers and collaborators in language understandable to them. The point of embeddedness, then, is that if science is going to be relevant to adaptive management, then it must be part of the process of discussing management goals and means. Hypotheses tested must be responsive to questions that affect the decisions at hand. Embedded science—science done by scientists working within a process—must be able to bring the scientific method to bear on uncertainties and disagreements blocking cooperative behavior.

Adaptive managers, ideally, would not, when faced with a problem of placing values on changes in natural systems, go to the economics department or the philosophy department to find a consultant to do a study at long distance. Rather, they should invite those practitioners—or local practitioners with similar skills—into the public process. If studies of attitudes and values become necessary in the ideal adaptive process, evaluation of possible development paths should be initiated by members of the management team. Alternatively, evaluators with special abilities may be brought onto the management team. The point here is that both adaptive science and evaluation of change is best done within the conflict-ridden context in which problems are faced, because each problem faced is unique.

CHAPTER 7

Tools of the Adaptive Trade

7.1. Static and Dynamic Deliberation: Two Functions of Tools of Evaluation

Adaptive management is all about living with uncertainty and responding to unexpected change. Can an adaptive process provide a means to question human values and preferences so as to create a rich deliberation about the appropriateness of current managerial goals as steps toward sustainable living? I think so, though this progress will require a discourse richer than that of economics. A normative discourse space in which objectives, goals, and social values are based only on fixed and preexisting preferences will be inadequate to generate the dynamism necessary to a full-fledged adaptive management process.

When estimates of WTP, or willingness-to-pay, are factored into a decision model, at least two important concerns must be raised. First, the argument of this book implies that it would be a mistake to think that *any* single model can comprehend all aspects of decisions in complex situations. While estimates of the current WTP of a population for some commodity might be helpful in making certain decisions, this approach cannot offer a comprehensive accounting of all benefits and costs. That accounting system, if its currency is WTP measures, must be recognized to be an incomplete, even incompletable system.

Second, while one can be very careful in developing methods for estimating WTP in complex situations, it must always be recognized that these estimates can at best provide a snapshot measure of present values. The language and practices of economic valuation, which embody a methodological commitment to reporting the preferences individuals express at a given time, is confined to the discourse space of static values and descriptive measures. Economic valuation asks the question, What values or preferences do members of a population have at this time? It is ill suited to ask the question, what values will or should a population express and pursue if its members wish to live sustainably (Norton, 1994; Norton et al. 1998)? To even begin to address this latter question, a more dynamic discourse is required.

Learning to be sustainable will require humility about current goals and values, as well as a recognition that goals and values will be reconsidered and reordered as needed. This learning requires an intellectual space in which people can learn about their values and those of others and can then reconsider the values of their evolving community. Environmental policy discourse, insofar as it can be useful in learning to live sustainably, requires a space in which a community can deliberate and learn what its members truly value and in which misguided goals and objectives are criticized and revised. Of course, all this discussion will have to be integrated into day-to-day practice as well as intellectual engagement. In this chapter, such a discourse—and a variety of tools to sharpen it—will be introduced.

In this dynamic intellectual space, problem formulation itself is recognized as being fraught with values: differing values account for the wickedness of many problems. So the tools offered will not provide recipes for correct decisions, and they will definitely not provide comprehensive, decisive principles applicable in all cases. Nevertheless, these tools can strengthen dynamic evaluative discourse. Pluralism, here, is an advantage; since the goal is "robust adaptation," what is needed is a variety of tools that can provide a web of partial considerations, from which one can fashion a strategy for evaluating change in particular situations. Unlike those pursuing a single right answer, who must build chains of reasoning with the strength of their argument at risk from the weakest link, I follow C. S. Peirce in seeking a "cable" of understandings, arguments, and reasoning rather than a chain of deductions.

Section 7.2 contains a survey of a few of the tools that can be effective in creating a learning path regarding normative issues, tools that function within the dynamic evaluative space defined above. Before presenting this catalog of tools, I close this section by highlighting the importance of the distinction between static and dynamic value discourses by examining one particularly popular analytic tool: the much-discussed concept of "ecosystem services." This concept has been used in a number of ways in public discourse, with some major studies focusing on the provision of ecosystem services as their central tool of measurement for understanding environmental values (Costanza et al., 1997; Daily, 1997; Gomez-Baggethun et al., 2010). Here I recognize that this tool has legitimate and useful applications, but it can also be argued that the concept is susceptible to confusion and counterproductive applications.

Distinguishing between the benign and not-so-benign uses of the concept of "ecosystem services" can be aided by examining the concept's possible roles in the two "spaces" defined above. Does discussion of ecosystem services contribute to the static or the dynamic space? In fact, it can be used in either way; a brief examination of these alternative roles in this chapter will shed light on its helpful uses and possible abuses.

Speaking historically, the idea of ecosystem services first surfaced in the late 1970s as a way to frame the benefits derived from ecosystems in utilitarian terms, based on the hope that such a framing would create greater public interest in protecting natural systems and biodiversity. During subsequent decades, the concept has entered the mainstream of environmental policy discussions and has provided the dominant framing for the much-cited *Millennium Ecosystem Assessment* (United Nations, 2005) and for a similar European Union report, *The Economics of Ecosystems and Biodiversity* (referred to as TEEB; Sukhdev et al., 2014). Given the difficulties in establishing protection of biodiversity and other productive biological systems, new arguments for protecting these resources are, of course, welcome.

Can the ecosystem services concept be used within the static space in which economists estimate and aggregate consumers' current WTP's? Should it be expected to "fill in the blanks" of an incomplete cost-benefit analysis (CBA) by providing economic information on values that cannot be or have not been measured within the mainstream of environmental economics?

Given how the concept was introduced to most advocates, researchers, and practitioners of public policy fields, it certainly seemed that its advocates were proposing a new and more comprehensive way to estimate economic impacts, one that would create broad support for conservation. In a multiauthored paper in *Nature*, the goal was stated as seeking the "total global value of ecosystem services" (Costanza et al., 1997). Associating uses of nature with various resource "sectors," the authors assigned value to sectoral processes and to the many services that humans derive from nature, then assigned economic values to services to humans that support or enhance human well-being. Authors of this paper surveyed the products and services humans derive from nature, placing dollar values on ecosystem services representing seventeen "ecosystem functions" such as climate regulation, water supply, soil formation, waste treatment, and recreation, and offered the eye-catching figure—"almost sure to be an underestimate"—of $16 to $54 trillion.

In the context at the time, many critics were criticizing economists for ignoring nonmarket values and arguing that there are important values not captured in standard CBAs, citing a variety of hard-to-measure values. Others argued that some values, like relationships and love, should not be subject to CBA even if they could be (Kelman, 1981; Sagoff, 2004). Whatever the intentions of the authors, and given the huge estimates of ecosystem services in the 1997 paper, it was tempting for readers and advocates to see ecosystem service estimates as a way to "fill the blanks" in what were viewed as incomplete CBAs or, perhaps, to replace CBAs with alternative, more comprehensive measures.

Note that, if one views ecosystem service concepts in this way, as a supplement to or replacement for incomplete CBAs, one must locate ecosystem service concepts and language within the static space in which dollars represent current preferences and preferences are fixed and unchanging—for that is the currency of WTP valuation. Unfortunately, there are three extremely powerful reasons to doubt that quantified ecosystem service concepts can provide numbers by which to finish or replace CBAs.

First—and this is a technical objection of great power—the dollar values measured in a CBA are *marginal* values; they are estimates of the price a given consumer will pay for the last item of a particular type chosen, so prices are constrained by the law of supply and demand and by the endowments of the consumer. The dollar values provided in the paper in *Nature*, however, represent the total values derived from certain services, divided by sector, and estimating the *total* value of a class of goods and services derived from that sector. These total values are often quantified as an estimate of replacement costs for the good or service if it had to be produced by human action. But since total values and replacement values are not set in an active market, such figures are usually overestimates because they do not allow real consumers to choose the next-favored alternative, but rather they list the entire cost of losing the service. The dollar figures involved in the two analyses are thus not commensurate. To add them together is to add apples with oranges; what one, in fact, gets is either nonsense or numbers that include a huge amount of double counting (Bockstael et al., 2000).

The second objection is that considering ecosystem service values rather than the more standard WTP values of mainstream economics does not ultimately gain much clarity over the problematic attempts to assign a dollar value to nonmarket as well as marketed goods in developing CBAs. Advocates of ecosystem service analyses list three general types of ecosystem services: provisioning (provision of food and fiber, and so on), regulation (maintenance of resilience, and the like), and cultural ecosystem services. The first two categories are generally handled pretty well through estimates of WTP, but when one tries to estimate values of "cultural" aspects of ecosystems services, all the problems that were encountered when economists try to estimate nonmarket values arise again.

The third objection is that, even though the methods of estimating effects on human welfare used by advocates of quantified measures of ecosystem service values differ from the methods of mainstream economists, they rely on the same basic metaphor or analogy. Economic value, measured in either way, relies on the metaphor of nature as a productive factory, so it should not be thought that introducing ecosystem service accounting enables analysts to explore a truly alternative notion of environmental value.

For these reasons, I conclude that introducing monetary estimates of ecosystem service into the discourse about nature and human well-being, and especially aggregating them with WTP values, is misleading and creates confusion. This is not necessarily to say that ecosystem service discussions, including discussions of monetary payments for such services, have no role at all in the static space of economic analysis. Monetization and payments for ecosystem services can be helpful in designing and implementing efforts at negotiation and compromise. The key difference here is one of context. If the context is one of seeking a totalistic (or as-complete-as-possible) estimate of costs and benefits, the above negative arguments apply. If, however, the context is one where there already exist groups and institutions devoted to finding cooperative solutions that improve the lot of all, then discussion of payments for services and attempts to estimate avoided costs may be very helpful in negotiations about how to fairly share costs of these improvements.

For example, in classic upstream-downstream discussions of river management, it may be useful to think concretely and monetarily about costs incurred by upstreamers and benefits passed to downstreamers if they agree to manage cooperatively. Here, the goal is not to set policy by aggregating total costs and total benefits of actions, but rather to find ways to encourage participation and to identify arrangements that are fair to all participants.

Indeed, there are real-world examples in which ecosystem service discussions and deliberations have contributed to greatly improved outcomes for all by virtue of transfers from downstream consumers of water to upstream users of riparian habitat. One such case, a water fund in Ecuador, has a long-enough track record to illustrate our point that using the concept of ecosystem services has aided in gaining cooperation and joint agreements acceptable to parties usually at odds. With Ecuador's large capital, Quito, perched on the western slope of the Andes, downstream of significant agricultural and other development activities, the potential exists for both conflict and cooperation between the area's upstream and downstream communities. Quito gets all its drinking water from rivers and streams descending the mountains, and that city was experiencing large and increasing costs to clean up water that had been degraded upstream by sediment, animal waste, and other pollutants.

Following a highly regarded project by which New York City reduced its water treatment costs significantly by protecting upstream stretches of the Hudson River with investment in improved agricultural and developmental practices, international and local Ecuadorian groups cooperated to create a fund based on initial capital from the Nature Conservancy and the US Agency for International Development (AID). These voluntary gifts to the fund were then used as an endowment to support both improved conservation practices among private landowners and improved management of the huge Condor Bioreserve that contains 5.4 million acres of protected lands in seven protected areas. The fund, which is renewed by transfers based on lower water treatment costs downstream, is then used to provide incentives to protect upstream watersheds by creating incentives for upstream users to reduce the downstream impacts of their activities.

This system clearly has an economic foundation and driver, but it also creates exchanges, including barter and in some cases payments for easements. The negotiations are overseen by an executive board composed of stakeholders from both upstream and downstream communities, which manages the trust fund with the help of a technical secretary, approving projects that shift incentive structures to protect in-stream water quality and address priority issues in protecting natural habitats and rare species.

The executive board and cooperating nonprofit organizations meet with upstream communities to determine what they need to improve their living situation, and they then use this information to design in-kind investments that will both protect water and improve the lives of residents in upstream communities. Representatives of upstream communities who are members of the executive board, provide ongoing feedback. Funding in this case and in many others involves a trust fund, with all or most of the money raised going to an endowment, so that projects can be paid for from interest on principal.

The executive board works with AID and NGOs as well as other institutions and organizations to develop projects that will protect water quality and also conserve ecosystems and biodiversity. For example, it offers fencing to farmers to keep their cattle out of streams and protect a natural buffer. In exchange for giving up some of their grazing land to fencing, the fund provides farmers with better-quality cattle that will provide the same amount of dairy products with fewer grazers. In addition, the fund will fence small areas on landowners' property and provide all the training and seeds necessary to grow vegetables for their family's use and perhaps some truck-garden items for sale. The farmers thus reduce food costs for their families by producing vegetables and may even find some ways to diversify income.

The first-instituted water fund in Ecuador—plus at least thirteen such funds either in operation or in planning stages in neighboring countries—are becoming increasingly important means to protect natural diversity while contributing to the well-being of upstream communities. The Nature Conservancy is careful to point out that the funds and the ad hoc arrangements they require in particular situations mean that each fund has its own structure, its own difficulties, and its own measures of success. Development of such funds, and variations on the theme, however, provide significant hope for developing community-based, stakeholder-managed adaptive management projects.

Notice that, in this example, the actual context of the static, economic model is adequate. The demand for clean drinking water is an unquestioned public good, so the question is how best to pay for it. When groups are ready to seek cooperative solutions, adopting an economic model of reasoning by which each group represents its own interests in dollar terms can create comparability and the possibility of negotiation and compromise. What is encouraging about cases like this is that the monetary estimates of ecosystem services are not part of some impossible dream of a complete and final CBA, but rather provide a set of tools that can be helpful to a community already engaged in negotiations. In these cases, a sort of exchange is taking place between upstream and downstream participants, and the static calculus of values is adequate to inform the groups' discussions. In these negotiations, the numbers may be very helpful in zeroing in on appropriate payments in the actual situation, by identifying connections between upstream practices and their impacts on downstream water quality. Such cases make legitimate use of the static language of microeconomics within a broader negotiation as a part of a public process.

In such contexts, where ongoing negotiations can be informed by estimates of ecosystem service values, these estimates function to make disagreements clearer and possible compromise solutions more obvious, and the negotiations are embedded in ongoing institutional arrangements. So, as long as the context invites self-interested negotiations among stakeholders trying to cooperate, payments for ecosystem services can contribute by clarifying tradeoffs and by purchasing buy-in from reluctant stakeholders. And all this can take place within the static space of economic analysis wherein preferences, as represented by the stakeholders, can be thought of as fixed and unchanging.

I conclude that ecosystem service concepts, viewed in a static mode, have a specific role within certain negotiation processes and that these concepts can be considered a useful tool in those specific contexts. What should be avoided, however, is treating ecosystem service values as providing methods that extend and somehow complete the process of aggregating all costs and all benefits of choices and activities.

7.2. A Peek into the Box of Dynamic Tools

The argument of this and the previous chapter has been premised on a separation—somewhat arbitrary, but nevertheless useful—of environmental discourse into two intellectual spaces, one "static" and the other "dynamic." When policy discussants incorporate language and assumptions of economics, viewpoints are expressed within a static space. In such an intellectual space, there is no room to explore, criticize, and possibly change people's preferences. If one sees the process of choosing goals and objectives as a dynamic one, as one element in an experimental, adaptive approach to management, we must move out of that static space and create an intellectual space in which public deliberation can lead to questioning, reformulating, and changing environmental values as espoused by a population that is in the process of learning to live sustainably. In this section, I explore some available tools that, if used within a good-faith deliberative process, can encourage the evolution of individual and group preferences toward goals that, over time, express themselves in more sustainable lifestyles.

7.2.1. Tools of Dynamic Analysis: A Survey

There is no denying that the CBA tool is a remarkable invention. By applying precise definitions and a carefully circumscribed method, economists can represent many values in the common currency of WTP. My complaint is not with the effort to use this precise tool in appropriate situations but rather with the reductionistic tendency of advocates (Economists) who wish to use the tool as a comprehensive accounting system for all environmental values. In this section, I survey the tools that are available to encourage, within an adaptive, collaborative process that pursues deliberation and learning, a dynamic assessment of environmental values, as listed in table 7.1. I will, in

Table 7.1. Dynamic/deliberative tools

A. Conceptual/analytic tools	B. Behavioral/action tools
1. Ecological footprinting	1. Scenario building
2. Safe minimum standard/PP	2. Participative modeling
3. Cost-effectiveness analysis	3. Back-casting
4. Hierarchy theory and systems analysis	4. Deliberative monetary valuation
5. Risk decision squares (a metatool)	5. Various group exercises (scientific/technical committees, citizens' juries, deliberative opinion polling)

this subsection, classify and summarize a range of tools available to aid in deliberative processes and, in the next subsection, I will focus in more detail on the implications of using scenario development and back-casting in conjunction with each other, because they illustrate additional general insights about the use of tools in dynamic evaluation. Here, though, I introduce a basic distinction between two kinds of dynamic tools: *conceptual and analytic tools*—ones that help describe and understand the system humans live within and manage—and *behavioral and action tools*—ones that prescribe activities that encourage dynamic evaluation.

These tools represent some of the resources available to individuals and groups that set out to learn by engaging an adaptive, collaborative process, and by addressing felt problems by engaging neighbors and cohorts to work toward a more sustainable future. Conceptual and analytic tools help understand and analyze a problem in new ways that may shift perspectives and priorities—what some authors call "double-loop learning" (Lee, 1994). In contrast, the list of behavioral and action tools provides a set of activities that can encourage building trust among participants and also framing questions and problems in new ways within an interactive process. These tools are discussed in detail in the following subsections.

7.2.1.1. Conceptual and Analytic Tools

A1. ECOLOGICAL FOOTPRINTING. Developed in the 1990s by William Rees and Mathis Wackernagel, ecological footprinting provides a useful way to understand and compare different activities by individuals and groups by calculating the human spatial demands placed on the earth and its surfaces to regenerate resources and provide ecological services (Wackernagel and Rees, 1998; Global Footprint Network, 2012). This versatile tool allows the comparison (for a given population) of the demands—measured in terms of land surface required for various activities—that that population places on natural systems. The tool, which gauges the spatial extent of the land and marine ecosystems necessary to support a given pattern of resource use, can be used to compare the consumption patterns of any group with the average consumption of other groups worldwide. It can also be used to compare a given level of consumptive use among some group with the average space available on earth, which in 2006 was calculated to be 1.8 global hectares per capita. This tool can be particularly effective because it allows groups both to compare their own performance against other international, national, and local groups and also to track the footprint of a population or group over time. A tool such as footprint analysis can be extremely useful in translating vague conceptions of human dependence on ecosystems into estimates (however tentative) of effects on ecosystems. It is surely not a perfect tool, though it can also function as a catalyst for concrete discussions of impacts on the ecophysical system, and it can provide some rough guidance as to the direction of greater degrees of sustainability of groups and their practices.

A2. THE SAFE MINIMUM STANDARD OF CONSERVATION (SMS) AND THE PRECAUTIONARY PRINCIPLE (PP). When a decision includes a great deal of uncertainty and a concern about irreversibility of impacts (as in the management of endangered species), and if any attempts to identify future benefits of a protective action are highly speculative, then some economists and other policy analysts favor using SMS for analyzing and making decisions. This decision rule states that, when protection of a valued ecological asset is uncertain and the benefits are difficult to quantify, then the resource should be saved, provided the social costs are bearable. This tool assumes wide agreement on a goal, such as the goal of protecting biodiversity generally or even every species in a given area, and then it urges that one pay attention to costs rather than benefits, seeking a solution that is "affordable" to the community (Ciriacy-Wantrup, 1963; Bishop, 1978; Norton, 1987; Berrens, 2001). In Europe and in some policy circles, PP is invoked. This principle, which urges caution when uncertainty and stakes of a decision are high, is at least similar to the SMS in that both shift the burden of proof onto those proposing risky policies or projects. One might prefer SMS because it is a bit more precise in that it focuses on bearable costs and emphasizes protection of resources even when benefits are difficult to count.

A3. COST-EFFECTIVENESS ANALYSIS (CEA). Many economists and others have noticed that, while attempts to provide comprehensive accounting of the benefits of a policy or practice are terribly difficult—indeed, we have argued that they cannot succeed—it can be much easier to estimate *costs* of a proposed activity. This approach often works well in combination with SMS. In situations in which there is wide agreement with a resource-protection goal and a community or government agrees that there is good reason to save the resource in question if the social costs are bearable, then it becomes a game of choosing the least costly method to achieve the given goal. Economists can be engaged to provide an analysis of the costs to society of various proposed protective procedures to achieve the agreed-on goal. Tools A2 and A3 can be used together when there is more agreement on goals than on means to achieve them. These rules can help avoid ideology and focus attention on agreed-on goals, such as the legally binding Endangered Species Act in the United States. Given a clear and legally binding law, protection of species can be taken as a given, and attention can turn to finding less-costly means to the agreed-on end. CEA provides an important role for economists in evaluating policies, even while avoiding the difficulties of providing comprehensive benefits analyses.

A4. HIERARCHY THEORY AND SYSTEMS THEORY. Current management thinking—indebted to John Muir and Aldo Leopold—is deeply committed to systems thinking and to more holistic and multiscaled structures of explanation. Over the last half-century, ecologists and others have developed multiple methods based in systems thinking. While some of these thinkers have been overly optimistic in seeking a single, correct system of representation, others—such as Werner Ulrich (2003)—have recognized that self-conscious attention to the boundaries and membranes attributed to systems under study can liberate thinking from single-level analysis and can thereby enrich understanding. There is no limit to what adaptive managers can learn from the exploration of systems in all their many representations.

Much has already been said about hierarchy theory (HT) and its usefulness in sorting out confusions of scale in both space and time, so the discussion here can simply highlight the uses of this tool. It designates a class of open-system models that are organized hierarchically, in that they focus on the world from local perspectives such that each level in a nested hierarchy is both a part on one level and a whole system on the lower level. Smaller, embedded systems change more rapidly against the backdrop of constraints imposed by larger, slower-changing systems. As noted, HT is a tool that helps in the construction of layered, multiscalar systems that can begin to represent the complexity of systems. This tool does not claim to provide a single, correct hierarchical model; rather, it provides many scale-sensitive models that can be developed or chosen to appropriately represent and relate the various elements of a complex problem that manifests at multiple levels.

These models can provide conceptualizations and formalizations that organize thinking about boundaries and scales of the systems that are being managed.

A5. RISK DECISION SQUARES (RDSS). This tool is actually a metatool in that it provides guidance in choosing appropriate decision rules in reacting to a given problem. The metatool relies on HT to provide general guidance as to which decision rule is appropriate in choosing a policy for a problem based on the ecological extent of possible impacts and the time of recovery of the system in worst-case scenarios regarding impacts of that policy. Since they may provide a useful metarule for choosing a decision rule appropriate to a given problem or decision, RDSs will be discussed in more detail in subsection 7.3, where the topic is how to live sustainably by choosing among multiple rules so as to apply the appropriate one to the decision at hand.

7.2.1.2. Behavioral and Action Tools

Whereas conceptual and analytic tools aid in understanding and deliberating about problems that a community might face within a collaborative process, behavioral and action tools provide activities that can improve communication and encourage deliberation about what to do and about which goals to pursue within an adaptive collaborative process. Here, I survey the activity menu briefly; some important implications and concerns about several of these processes and their interactions will be further addressed in section 7.2.2.

B1. SCENARIO DEVELOPMENT. In many cases, groups seeking to identify and pursue a sustainable pathway to the future engage in developing scenarios that characterize multiple possible futures for their community. This tool functions well in conjunction with the idea of evaluating development paths. Scenario analysis, originally developed for military purposes and later modified for use in industrial planning by Royal Dutch Shell, provides communities that undertake such an activity with an opportunity to explore possible futures. As such, scenario analysis can be a key component of any evaluation process that hopes to choose among possible pathways into the future for a given community. Usually working with a facilitator, groups or entire communities propose possible pathways into the future, and through a discussion process, the group culls and combines these pathways into a small number of possible futures. Then, associating these pathways loosely with various policy options, participants begin to think in terms of preferred outcomes and to weigh how various policy choices encourage trends that may contribute to arriving at favored outcomes. Scenario analysis can be used in many, diverse contexts, and sometimes in conjunction with tools B2 and B3, which provide some technical tools to enhance the scenario-building process.

Scenario analysis has been widely used in city and industrial planning as well as environmental policy discussions. This technique has been usefully adapted to environmental policy discourse and planning. Successful application of this technique typically requires the existence or creation of a set of interested observers; ideally these would be representatives of the parties affected by a decision or proposed policies, whose participation would be on-going as part of an iterative process of deliberation and information-seeking.

When participants in a process decide to undertake a scenario development exercise, usually one or more facilitators are chosen, who then plan and manage a process of several steps, with the possibility of open-ended adjustment and rethinking of the process and the scenarios at each step. The first step involves endorsing a set of assumptions about the current situation and the identification of certain drivers that can be expected to create bifurcation points in the future. Either through informal "brainstorming" or through a more formal process of participatory modeling, the participants then propose and sketch out a number of miniscenarios.

These miniscenarios, preferably numbering less than ten or so, are then reduced by combining features to a smaller number—usually two to four scenarios. Reducing the number of scenarios is important because experience shows that dealing with a larger number of scenarios is confusing and, in the face of too many of them, policy-making users will tend to fasten on only one, favored scenario at the expense of the alternatives. Once the scenarios are reduced to a manageable number, they are often given catchy names and written up as a narrative that might be told by looking at present and subsequent events from a hypothetical future. Since the events and points of change will be related to possible policies that might be instigated, analysis of the scenarios can help groups to focus on possible trends and to differentiate different possible paths into the future. By envisioning different pathways into the future, scenario analysis can be useful in highlighting possible negative and positive outcomes and by identifying objectives that could be pursued through appropriate policies. Experience suggests that it is important that groups work with scenarios with time horizons of more than ten years to avoid simply projecting present trends and shifting focus to slower interactions of trends that in the short term seem independent.

In general, scenarios should be distinguished from predictions as to what will happen. It is better to think of them as "projections" of possible trends into the future rather than as concrete expectations as to what will, in fact, happen. What seems to be important in practice is the experience of connecting possible choices and policies with trends and with the clarification of values. Participants generally note that "messing around" with the scenarios deepens their understanding of choices and helps develop possible objectives for management.

B2. PARTICIPATORY MODELING. As conservation efforts are more and more directed at landscape scales, it is sometimes useful to engage in participatory modeling in an attempt to understand system functioning and achieve a balance between conservation and development goals. The idea is to create a model that represents the complex system involved in a given management problem. In some cases, collaborative groups engage in long-term modeling efforts, initially creating a model and then revisiting it in subsequent periodic meetings. Such models can be invaluable if they attain the status of a boundary object, which increases communication across conceptual barriers. Building such a model can be an important element in an ongoing, iterative process of developing a better understanding of the systems involved and the options available to manage them. More often, groups engage in what has been called "throwaway modeling" in which stakeholders are brought together for a short (one- to three-day) exercise in which policy makers, scientists, and laypersons work to create a hybrid model that is informed by available science but simplified and modified to focus attention on problem drivers and management options.

Here, the process of discussing, constructing, and refining the model is considered more important than the model itself, especially with respect to throwaway models but also with respect to more iterative ones (Sandker et al., 2010). In the cases of both iterative modeling and throwaway modeling, advocates argue that such exercises improve communication among stakeholders and groups with differing interests, and there is evidence that these processes increase understanding and agreement about the functioning of complex systems (Rouwette et al., 2002).

Participatory modeling can contribute to a more sophisticated understanding of problems, create common understandings among groups that begin with very different perspectives, and improve trust and effective communication. As such, participatory modeling can provide a tool that stimulates rethinking of both empirical and evaluative viewpoints and contributes to a dynamic process of social learning.

B3. BACK-CASTING. Scenario development and analysis is sometimes supplemented with an associated practice called "back-casting" (Robinson,

1996). This practice involves, first, developing one or more scenarios out for several decades, say fifty years. If a given scenario emerges as a favorite of the group, the description of the outcome of that scenario in fifty years can be thought of as a rough characterization of a set of objectives that might guide the practices of the society in pursuit of those objectives. Back-casting, then, asks a series of questions based on the desired outcome so characterized; let me call it "the desired future state." If one wishes to be at the desired future state in fifty years, where would one have to be in thirty years? If one wishes to be on track for fifty-year goals in thirty years, what would have to be accomplished within twenty? If one hopes to be on track at twenty years, where must one be in ten? And, finally, if the hope is to have reached that stage in ten years, what needs to be done today? In this way, scenario development plus back-casting can provide valuable tools for exploring long-term values and for working one's way back from distant objectives toward today's choices and actions.

B4. DELIBERATIVE MONETARY VALUATION (DMV). As argued in a 2008 review article, Clive Spash has noted rapid expansion of research—in several disciplines—employing a combination of contingent valuation techniques developed by economists to estimate individual preferences, while also using deliberative methods generally developed by other social scientists. More and more researchers have been experimenting with augmenting questionnaire and other traditional "stated preference" data-gathering techniques with focus groups and other ways to involve respondents in discussion and clarification before committing to particular responses (Sagoff, 1998; Spash, 2008). Most economists have mainly seen the application of deliberative processes as a way to improve the validity of the stated preference techniques and procedures they themselves have used for decades to estimate and aggregate individual preferences. As such, this work does not go beyond mainstream economic theory and the practice of representing fixed preferences as WTP. Insofar as such studies are understood as contributors to a more complete and or accurate CBA, as the use of the WTP measure would suggest, they are confined to the static intellectual space of mainstream economics—a space in which preferences precede choices and preferences do not change. So understood, this trend is of little interest to us in our search for more dynamic tools to help us to learn our way toward sustainability.

Yet there is another trend also described by Spash (2008), and that is to link—both in research and in experiments in involving the public and stakeholders—monetary valuation exercises with techniques developed in the other social sciences. Unfortunately, results of these studies and experiments, even when they are designed simply to articulate a group-legitimated WTP and to validate contingent valuation-derived WTP figures, have not gone as economists had expected. Hoping to end the processes with a WTP number that is legitimated by group consensus, instead researchers have found that the deliberative groups often object to elements of the process, insisting on side qualifications and generally refusing to act like well-behaved members of *homo economicus*. They react, one might generalize, as pluralists who refuse to have their thoughts and feelings forced into the straitjacket of economic monism.

Spash (2008) summarizes his findings as follows:

> Reviewing empirical DMV studies evidences a range of issues regarded as external to economics and the validity of its methods, issues which are typically kept at arm's length by most environmental economists, namely multiple values, incommensurability and lexicographic preferences, social justice, fairness, and nonhuman values. . . . In other words, once one engages individuals on the complex and interesting subject of environmental values and environmental objectives, one cannot confine them and their discourse to the static space controlled by the static assumptions of economic valuation. (Spash, 2008, p. 469)

Faced with this empirical evidence, some economists and other social scientists have begun experimenting with alternative ways to articulate tradeoff decisions under deliberation (Aldred, 2001; Howarth and Wilson, 2006). What may be emerging is what was envisioned by Sagoff (1998) in one of the earliest experiments using monetary valuation together with focus group deliberation. Placing dollar values on environmental change can be a useful way to frame questions for more detailed discussion. Group deliberation then allows participants to learn from others and to refine their views, ending with a number thought by most participants to be more reasonable than the point at which they started. Such processes, either when the participants are pushed to come up with a dollar figure or when they are allowed to sprout new ideas and to express values way outside the monetary rubric, require an intellectual space in which values are expressed in multiple terminologies and discussions of value go much deeper than merely eliciting an individual WTP.

But one should not see moves in this direction as turning away from dynamic valuation—to the extent that deliberation brings up new ideas and new reasons, these exercises can be the spark that will ignite change in values and preferences. Given that dynamic valuation can contribute this function within an adaptive management process that is open-ended and seeking corrigibility for all beliefs and values, experimental and hybrid approaches from economics and other social sciences should be seen as tools by which communities can struggle toward values and objectives that are sustainable.

B5. VARIOUS GROUP EXERCISES. Groups involved in collaborative management efforts can also make use of a variety of side exercises, including the use of technical committees, focus groups, citizen "juries," and participative polling. Space does not permit detailed explanation of each of these activities, but each of them can be useful in some of the infinitely varied situations faced in group processes. Technical committees are given a charge—to survey and summarize the current state of knowledge important to understand crucial aspects of the functioning of the system under management. Examples of important contributions of technical committees will be detailed in chapter 9. Focus groups can be formed to deliberate about particular issues and controversies; small-group deliberations like these can either discuss values and objectives in general terms or, as Sagoff used them in section B.5, they can be used to reconsider and alter the initial statements regarding individuals' WTP for various improvements. In either case, they can function as catalysts for reconsidering objectives and priorities.

In cases where opinion among policy makers and participants is sharply divided over a well-defined issue within a collaborative process, formation of citizens' juries can address this problem through a small-group deliberative process. In such processes, a group of representative citizens (often twelve to sixteen individuals) is chosen. Time is spent bringing the group up to speed with respect to the issues, followed by presentations by experts and strongly interested parties. When all this information has been presented, the jury deliberates and develops a report, which is then entered into the debate as one informed group-opinion, in the hope that the report will lead toward consensus on how to address the problem (The Citizen's Handbook, 2013).

Another approach that can help sample opinion as part of a dynamic process that includes opportunities for social learning is the use of deliberative polling. Originally developed by James Fishkin, deliberative polling uses media (cable television or Internet) to create a forum in which representative individuals participate in a public process of information-gathering and opinion expression (Fishkin, 1991). Much like citizens' juries, deliberative polling attempts to provide sufficient information to members of the public so that more informed opinions can be expressed and reported to the rest of the public. An advantage of deliberative polling is that use of media can allow formation of much larger representative groups, which help ensure that all opinions are represented, providing a broader representation that improves both visibility and plausibility of the poll results. While complex to administer, this activity can be useful in expanding the base of interested parties and raising the level of understanding of complex problems among the general population.

7.2.2. The Dynamics of Objective-Setting

Having surveyed some of the tools available to catalyze a more dynamic approach to evaluating change, I turn now to a clarification of an issue that might be raised by some of these methods, especially when scenario development is combined with back-casting. Back-casting assumes that, at some point in the exercise of scenario development, a preferred scenario will emerge. The objectives embodied in the preferred scenario can then be worked backward to focus on more near-term activities that are conducive to achieving the future that is embodied in the objectives of the preferred long-term scenario.

But now the attentive reader may say, "Aha!" The story of this book has thus far been one of how to avoid Optim's slavish commitment to objectives and to a single vision of sustainability. Yet the scenario development exercise asks deliberative groups to project a vision some fifty years into the future. How does this differ from the strategy of Optim, who begins by stating objectives for the future—a sustainable outcome—and ends by reducing the problem of living sustainably to a matter of finding means to the optimal outcome? Does advocating the scenario-analysis-with-back-casting tool revert to Optim's strategy—the very one I set out to avoid?

I answer this question negatively, based on two clarifications. First, the optimization strategy does not merely require the prior statement of a set of objectives to be pursued for the future: the strategy also requires reduction of the number of objectives to as small a number as possible, ideally only one. So, the first reaction to this charge that I have retreated into Optim's strategy is to emphasize pluralism; recognizing a plurality of values will encourage the development of multicriteria management systems that avoid narrow optimization of a single variable.

A second, more important, response to the fear that use of scenarios and back-casting may cause my approach to converge with optimization strategies requires a distinction between two ways of thinking about objectives. Remember that, in our key distinction between substantive and process rationality, the former requires the specification of final objectives. Under Optim's understanding, objectives are set early in a process of managing for sustainability, and these objectives are then solidified to the point that, in ensuing discussions and choices, deliberation will be about the most efficient means to an unquestioned end—the objective chosen and the associated variable that is to be optimized. It is this aspect of the optimization strategy that makes it attractive as a means to reduce problems to technical issues addressable by experts and technocrats. I refer to this way of understanding and positing sustainability objectives as the establishment of "hard" objectives. Hard objectives are objectives that, once set, are immune from criticism, at least as far as the optimization strategists are concerned.

But objectives can also be understood as "soft objectives." Soft objectives are conditionally accepted, always open for criticism and modification. Indeed, the kind of objectives that will be most useful in an adaptive management process should be formulated in such a way as to invite further refinement; one might even say that the soft objectives most useful in adaptive management should be adopted temporarily, with future questioning and reconsideration built into their specification. This is, as one would expect, an ongoing, iterative process. Identifying and pursuing soft objectives are, in fact, essential if adaptive management is to encourage changes in goals, objectives, and values as new information comes in. One should not assume that these changes will be easy. Evidence shows that individuals, faced with arguments that they should shift goals, may trigger resistance because giving up old goals involves waste of sunk costs and also because many participants will simply be reluctant to change. It is only by expanding experimentalism to include norms, goals, and objectives that adaptive management can function effectively within the dynamic space necessary for the evolution of newly formulated goals and objectives.

This distinction between soft and hard objectives, while introduced here to explain the role of scenario development and analysis as a policy tool, will be applicable to other tools as well. In general, full-blown adaptive management as advocated in this book submits all disagreements to experience, and this process must apply to disagreements about objectives as well as about means to those objectives. The tools to be used in adaptive management as discussed here will involve the setting of goals and objectives to undertake experimental actions. In contrast to the "hard" understanding of objectives, Adapt suggests soft objectives that are understood not as unquestioned but as hypotheses open to testing by experience. This latter view places the tool of scenario building within an open-ended, adaptive process in which social learning about objectives and values is possible. (See figure 5.2 and National Research Council, 1996.)

Scenario-building on this adaptive view is not an exercise in setting hard objectives to reduce management decisions to questions of means only; it is a tool of learning in the broadest sense of that word. When this kind of profound learning occurs, priorities shift and differing concerns come to the foreground. Such learning will be necessary if today's societies are to learn their way toward sustainability.

7.3. Living and Flourishing with Many Tools

As hammered home in part 1 of this book, there never has been—*and never will be*—a complete accounting of all environmental values favored by members of a population. Economic analysis has the significant advantage that it offers a method for estimating current preferences. Most people, however, when invited to express their environmental values in terms of dollars, will go along if the defined commodity sounds like something they might purchase in a market, but will protest in various ways when asked to quantify their value for items quite unlike commodities, such as friendship or even "ecological resilience."

Looking again at the list of tools in the previous subsection, notice that most of them provide techniques and tricks to encourage a broad, dynamic discourse about environmental values. Looking a little deeper, note that each of these tools exploits or develops an analogy or metaphor. A scenario is like a short little play, perhaps within a larger story, that tells its own little story; so scenario analysis suggests that deliberators tell stories as a way of clarifying what is important to them as events unfold in the future. Ecological footprint analysis employs a spatial metaphor, plotting human uses of nature onto a spatial grid of finite space. Sagoff and Spash both suggest that groups evaluating policies might learn by trying to apply the market metaphor, taking into account costs of negative externalities and, by listening to the viewpoints of others, adjusting prices to obtain more thoughtful estimates of fair prices. Citizen juries, of course, exploit an analogy with the court system's effective approaches to evaluating evidence and reaching conclusions. The tools available to engage in dynamic environmental evaluation are many, each being adaptable to multiple situations, and each may be exquisitely appropriate in one situation and worthless in other situations. Choices of tools are generally guided by metaphorical thinking.

So how should we react to this plethora of tools and associated metaphors? At first, it may seem that the choice of metaphors and analogies, and the tools associated with them, will be chosen arbitrarily; every attempted clarification of a situation might be thought to be idiosyncratic. Are pluralism of values, pluralism of tools, and pluralism of metaphors merely recipes for chaos in environmental policy discourse?

Not necessarily. Plurality of tools can actually be an advantage when facing complex and controversial situations. Environmental problems are arrayed across multiple scales, and the formal tool, HT, can be useful in choosing the most appropriate spatiotemporal scale to address and illuminate a given situation. Because different environmental problems emerge on different scales of space, with different temporal horizons, it is often the case that different problems are associated with different physical and ecological dynamics (Norton and Ulanowicz, 1992). What was learned from Leopold's simile of thinking like a mountain was that the dynamics of deer populations unfold on at least two scales. On one scale, wolves and hunters will be major factors in determining the crop of deer in the next hunting season. What Leopold later realized was that a much slower dynamic, an ecological and physical one that creates topsoil and allows the growth and availability of browse for those deer, was affected by his drastic action of completely removing wolves from the system. Killing wolves appeared to be a reasonable policy when considering only the short-term dynamic of wolves, hunters, and deer. As it turned out, Leopold learned a hard lesson: by paying attention only to a short-term dynamic, he created a disaster that ensued because his actions in removing all wolves affected a larger-scale and normally slower-changing variable—the development and maintenance of soil and vegetative cover.

One way to better understand Leopold's mistake—not an unusual one among environmental actors who create tragedies of the commons—is to say that he applied the wrong decision tool because he did not understand that his actions had adverse long-term consequences. To apply what we have learned from Leopold's mistake, it is useful to introduce a bit of an idealization: let us divide the process of making important environmental decisions into two "phases." It is an idealization because, while the label "phase" suggests different stages in a series, in practice both phases often, even usually, are active at the same time. I call the two phases (1) the action phase and (2) the deliberative phase. Because Leopold rushed into action without considering broader and temporally longer consequences, one can say that he (not unlike most humans) acted without sufficient deliberation about longer-term effects. Since he learned a great deal from his error, we will find that many of the hard lessons we learn in a deliberative phase result from actions and our experiences following those actions. Actions, when thus viewed retrospectively and in practice, are often the occasion for learning in the deliberative phase. So, while the idealization may seem to sequence deliberation before action, in fact the two interact with each other as they both unfold through time, with actions based on too little deliberation still teaching important lessons.

Spatial Scale of Impact

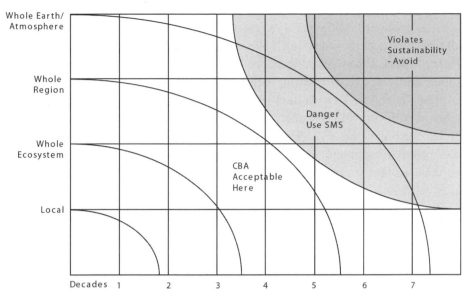

Figure 7.1. Risk decision squares: The squares define a "decision space" that corresponds to the spatiotemporal features of a problem. Decisions range from easily reversible to irreversible, measured as the time required for impacts of the decision to be reversed. As one moves across the horizontal axis, impacts of a given decision are more and more difficult to reverse. Decisions with impacts over larger and larger areas are plotted on the vertical axis, representing increasing spatial extent of risks. Plotting these two variables can characterize decisions according to the spatial extent of their effects and the time required for recovery from the impacts of that decision. RDSs can thus be used informally to provide a basis for choosing the proper decision rule in various situations (modified from Norton and Toman, 1997).

Turning attention to the deliberative phase viewed prospectively, we can ask whether it is possible to choose, given that there are multiple tools available, the most appropriate tool for a given problem. In the deliberative phase, the focus is on choosing an appropriate tool for the analysis and understanding of a problem. And, given an emphasis on the scale of problems—and the opportunities available—one has to speculate about what are likely and possible outcomes of a proposed action, using HT as a metatool to "locate" a problem in a decision space we can call a "risk decision square" (RDS; see figure 7.1). An RDS can function as a metatool in the deliberative phase to, first, classify a proposed action according to its possible effects at different scales of space and time and, second, to use this classification of problems according to their scale and the scale of the dynamics they affect to choose an appropriate tool for the situation.

RDSs exploit the relationship posited by HT between temporal and spa—

tial scale to plot problems along the two dimensions: temporal reversibility (how long will it take for impacts of an action to correct themselves if effects are unacceptable?) and spatial extent of effects (how large an area will be affected by the proposed action?; Norton and Ulanowicz, 1992; Norton and Toman, 1997). Accordingly, decisions that are easily and quickly reversed can be addressed by comparing costs with benefits, based on current preferences. Or, if a decision is likely to have effects over very small areas, or affecting only one of several similar systems, decisions can be based on an accounting of local preferences. And, trivially, if any decision has a likely short-term impact over a small space, fulfilling both conditions for applying a static calculation of benefits and preferences, a CBA will be an adequate tool.

Placing decisions within the space helps one decide which of several decision rules should be applied. A problem that requires more than a human lifetime (say, seventy years) for recovery, in the case of a bad outcome, becomes one of judging sustainability. As one approaches the upper-right corner of the decision space, some decisions will require caution and some will be judged unsustainable. RDSs can thus be used, not formally but instead informally to provide a basis for choosing the proper decision rule in various situations (modified from Norton and Toman, 1997).

In Leopold's case, the removal of wolves from the whole southwest territories—a virtually irreversible action—should have, in the deliberative phase, produced much longer-term effects on large-scale ecological systems, and he should not have been so willing to trade off ecological harmony for a quick but temporary increase in the deer herd. He should have thought like a many-scaled mountain. Similarly, someone contemplating changes over a whole ecological system might, in the deliberative phase, do an ecological footprint analysis and recognize that the spatial pervasiveness of an impact calls its wisdom into question.

The recognition that there are multiple tools allows an open-ended discourse not only about what should be done but also about how to choose tools that are appropriate, given the spatial and temporal scale of likely effects of a given action. It is in this sense that an RDS can function as a metatool exploiting the correlation between temporal and spatial scales and can aid in determining which problems and actions—including no actions— have sustainability effects. When a problem falls in the upper-right corner, sustainability issues are clearly involved.

Most tools actually represent technical exercises that are based on underlying analogies and metaphors. In the deliberative phase, then, a major aspect of discussions—from characterizing the nature of a problem or a probable impact to choosing or advocating a particular tool, a major aspect of people's thinking and discussion will be both metaphorical and analogical—a subject to which we return in chapter 8.

7.4. Tools, Heuristics, and Transformatives: A Messy Workshop for Messy Problems

In this book, with the help of Adapt, we have populated the process-oriented path with quite a number of aids to deliberation—means by which participants in a process can adjust their objectives and learn to use the insights of science and technology more effectively. For clarity, we can divide our procedural aids into three categories: (1) tools, (2) heuristics, and (3) transformatives.

1. *Tools*, as explained in sections 7.2 and 7.3, are activities or exercises that can serve a stated purpose. They function best when the objectives in question are clear. Here, one can include the "tools" of contingent valuation and stated preference methods as well as revealed preference methods as means of estimating existing preferences. Due to the static assumptions necessary to use these tools in an effort to determine aggregated individual preferences at a given point in time, however, these tools are only useful in the static discussion in which individual preferences are assumed to be present-time snapshots. These tools can be useful in choices regarding where to invest in public goods like parks, and they have also proved useful in negotiations between upstream and downstream users of a river, once an ongoing process of cooperation has begun.

   Given an emphasis on social learning, one can emphasize another deliberative space, one in which the expressions of value are plural and in which participants learn by interaction with others and through adaptive experimentation. Here, given a particular situation, one can recommend tools like scenario development as an aid in moving toward consensus on soft objectives, the use of ecological footprinting to translate environmental concerns into spatial terms, or arrange a citizens' jury to resolve a particular disagreement. I hypothesize that these tools, if appropriately used as guides to effective deliberative processes, can encourage social learning by serving the goal of better communication and more cooperative behavior. I have thus emphasized tools that will serve the dynamic aspects of a truly adaptive process in which goals and objectives, as well as scientific uncertainty, are approached experimentally.

2. *Heuristics* are not quite like tools, which are directed at particular goals;

heuristics are more process-oriented. Throughout this book, Adapt has furnished "heuristics." They provide guidance that encourages healthy processes. Heuristics, then, should be read as rules of thumb that tell us what has worked well in similar and varied situations. Their application requires wisdom, especially in finding a reasonable balance among competing objectives and sensitivity to local situations, but the heuristics can help participants in an adaptive process to focus on more productive questions and to avoid some dead ends in negotiations. Wisdom seeks robust approaches and flexible policies, and it differs from optimization, which emphasizes one or a few variables that lock in hard objectives.

3. "*Transformatives*" is a term I have coined to stand for linguistic and cognitive tropes that can cause changes in perception and perspective regarding a situation or problem. Transformatives differ from tools. Tools are used to pursue a known or at least aspired goal, whether substantive or procedural. Transformatives must be capable of causing significant shifts in values and worldview. Most effective transformatives either involve the use of fresh or newly combined metaphors, or they exploit analogies not yet grasped. By introducing a new metaphor, one launches an alternative narrative, an alternative way of making sense of a complex world. Transformatives thus function in the dynamic, deliberative space in which values and preferences change, and include metaphors, similes, analogies, and ironies.

Of course, one could write a book on the differences and nuances among these tropes and devices; but here only one aspect of all of them is important. What they share is an ability to cause cognitive shifts from one perspective to another, which, in turn, can lead to value readjustment. The devices involved can perhaps be characterized as catalysts for restructuring reality and, of course, involve attempts to change behavior through structural as well as cognitive means, as recommended by Heberlein. I therefore—while apologizing for lumping them together despite differences—refer to these catalysts as "transformatives." The next chapter will explore the special role of transformatives such as metaphors and analogies in public discourse and in adaptive management. Transformatives are vital in generating "double loop learning"—the kind that causes changes in perspectives, viewpoints, and values.

Before closing this chapter, however, I turn once again to the popular concept of ecosystem services. Clarifying different roles it might play in evaluation of environmental change will illustrate the positive uses of the concept in more dynamic situations.

Above, I argued that identifying certain "services" and talking about how much it would cost to provide those services in artificial ways can clarify issues and encourage compromise and buy-in for public goods projects. Payments for economic services can also be useful within negotiation processes to help participants find fair compensation packages for groups whose resources have been sacrificed to provide an important public good.

But it was also argued that the static assumptions necessary to ascertain individual preferences at a given point in time disable participants from actively questioning their preferences and values. A more dynamic deliberative space is necessary if adaptive management and social learning are to shift perspectives. Can the ecosystem service concept serve an important purpose in a dynamic discourse capable of causing shifts in perspectives and values? Note first that any reference to ecosystem services is itself metaphorical. To discuss nature as providing "services" can broaden the usual categories of items taken from nature as natural products or as commodities made from natural resources. So, one advantage of the term "services" is that it refers not simply to physical products but also to values derived from processes like the regulation of systems for resilience. Just as the term "service industries" expanded the usual idea of industry as manufacturing, showing that many jobs do not produce some physical product, the term "ecosystem services" can help shift terminology and attention away from objects and toward processes.

Thinking of ecosystem services as an open-ended metaphor that encourages the public to think of nature as productive of more than just physical products can stimulate a change in what gets valued. The concept may encourage the public to see nature as providing resilience and other means to respond to change. When considered metaphorically, speaking of ecosystem services encourages people to think of a broader range of benefits than benefits to them in the form of material products. By encouraging individuals to ask, "What processes of nature are responsible for my enjoying X?" the ecosystem service conceptualization encourages discussants to find many new and broader ways that nature serves human interests. Ecosystem services can also be seen to serve local, aesthetic, and cultural values, so in general the idea of ecosystem services can serve as a catalyst for recognition of values that have been obscured by our overemphasis on the products (commodities) in many economic discussions.

Having said that, I also note that one effect of the rapidly growing popularity of the ecosystem service idea so far has been to increase the role of economic thinking in public discourse; witness the almost-exclusive emphasis on economic aspects of services in The Millennial Assessment and in TEEB

(United Nations, 2005; Sukhdev et al., 2014). Why is this? It is because ecosystem services employ the same analogy or metaphor as standard market economics does: nature functions like a large factory, producing goods for humans to consume and to support human welfare that is pursued in markets. Adding that the factory also produces "services" as well as "products" does not change the instrumental role that nature is given. The ecosystem service concept, which evokes the same "factory" metaphor that was evident in an economy focused on "commodities" and physical products, goes only a short way toward the development of more dynamic evaluations.

I have highlighted the role that the ecosystem service concept can play in encouraging individuals to trace benefits they enjoy back to ecological processes that may lead those individuals into a new perspective for highlighting new and more appropriate values. Is it wise to put all our analogical understandings into a single metaphorical basket? In the next chapter, I delve more deeply into the role of transformatives such as metaphors and analogies in forming and evolving environmental values within an adaptive management process.

CHAPTER 8

Constructing Sustainability: Imagining through Metaphors

8.1. A Single Metaphor?

Monism, as well as the ideas that support it, including the commitment to the idea of a single right answer derived from a comprehensive method, has many facets. One of these is the pervasive use of just one metaphor, that of choices in markets to understand all human decision making. Indeed, there may be little difference between the following:

1. The claims of Economism that all values can or should be represented as economic (WTP) values, and
2. The claim that we can or should understand human–nature interactions using the single metaphor of market behavior, with nature playing the role of a productive factory turning out raw materials for the production of commodities.

Moral monists approach ethics with confidence that there exists a central principle that yields rules that can, in turn, be applied to all situations with moral significance. Moral action on this view is action that conforms to the dominant rule—whether the rule is stated in terms of consequences or in terms of rights and duties. While it can be useful as an exercise to try to systematize moral thinking in this way, thus minimizing the number of central principles necessary to support all our moral judgments, the assumptions that support this principles-and-rules approach as a reasoned philosophy are, at best, heroic failures. The goal they use as their ideal and guide— that of eventually resolving all moral quandaries by application of a single principle—requires huge amounts of faith, optimism, and hand waving over gaps in the application of general rules. Indeed, most moral reasoning begins not with choosing general principles on the way to deducing rules applicable to a given situation but with reasoning about what kind of situation one faces—and that act depends more on finding an appropriate metaphor or analogy than on applying a rule derived from a general principle.

Pluralists should accordingly be pluralists with respect to employing guiding metaphors as well as linguistic expressions and varied articulations of value. The assumption of monism has led to an interminable debate over what is the correct principle to guide ethics. In many environmental policy discussions, there is one dominant metaphor, attributed to Garrett Hardin: the tragedy of the commons (1968). Hardin applied a single principle—that humans will always act to maximize their own (perceived) self-interest. Based on this principle, he argued that common resources such as a shared pasture will inevitably be destroyed because each herder will add more and more animals until the pasture is ruined. The herders—bound by the principle of self-interest—will reason that an added animal will increase his or her own herd, while costs in the form of damage to the pasture (and the expense to rehabilitate it) will be borne by the community. This metaphor of a commons tragedy illustrates how, for economists, the principle of self-interest and the corollary that environmental evaluation must be measured in terms of individual interest. This metaphor appears to one side in the debate as an argument for expanding private property and to the other side as an argument for more government regulation (Norton, 2005).[1]

Economic monism is embodied in Hardin's metaphor. In many situations the monism of Economism dominates discussion and provides an example of how a dominant metaphor can lock discourse into a single way of thinking. This classic metaphorical understanding of resource use stands in stark contrast with those who reject Economism in favor of a theory that all ultimate value is intrinsic value. These two competing monistic theories underlie much of the polarization and gridlock in environmental policy.

Traditional monistic systems of ethics understand moral action as mainly a task of *deducing* rules from central principles and then *applying* those rules to particular cases. Pluralism, by contrast, places greater emphasis on the empirical conditions of the situation and considers multiple possible metaphors and analogies, letting each of them highlight particular social values. Attention can then be focused on finding agreements that are consistent with multiple visions and characterizations of the problems. As these multiple visions and voices are expressed and brought to bear on the situation at hand, actions may be found that can be supported by coalitions, because such actions achieve an acceptable balance in supporting competing values.

Ostrom (2007) has shown that, if one respects the ability of members of communities to transcend their individual interests, then there are alternatives to the metaphors of tragedy for public goods. If one did not need to be simply pro or con with respect to one dominant metaphor, and if one could encourage pluralism and social learning regarding environmental values, then one could explore environmental problems from multiple metaphorical perspectives.

To consider the possibilities of developing a metaphorically pluralistic approach to value exploration, I return to the concept of ecosystem services one more time. I have argued that this concept, despite some dangers and limitations in its uses, has created a useful discussion of ways in which humans depend on natural processes. Even when employed within the static constraints of an economically defined intellectual space, the concept can improve communication and encourage both participation and buy-in as communities negotiate to improve the common good.

The ecosystem service concept can also contribute to dynamic discourse when it is noted that it is itself a metaphor. What one gets from nature is "like" services that support our well-being. So understood, ecosystem services, in addition to offering "provisioning" benefits, also support a strong value for natural regulatory processes such as resilience and health of ecosystems.

Further, the ecosystem services concept may be understood as an open-ended metaphor, one that can encourage individuals to connect some benefit they enjoy with specific aspects or characteristics of a natural ecosystem. In this way, ecosystem services can contribute significantly to the discourse about what is valued in nature, and why. Making these connections encourages normative sustainability judgments by connecting opportunities valued by individuals and groups with actual physical processes, thereby giving substance to the requirement that sustainability necessitates the specification of "stuff." Given these advantages of the ecosystem service metaphor, one might ask whether it will be sufficient—if developed and widely applied—to lead to a sustainable future, or to lead close to it. To this question the answer, unfortunately, is "No."

In the previous section, tools and transformatives were distinguished. A tool can be associated with a desired goal or objective, whether that objective is soft (open for experimentation and alteration) or hard (an anchoring optimum to be pursued without further deliberation). Tools are not transformative, because they are constituted by their contribution to solving known problems in the service of goals, including those of communication and cooperation. But transformatives, which include metaphors, analogies, "models," and similes, do have the power to change the way one sees the world; they can, as Leopold learned while contemplating his "newly wolfless mountain," alter one's point of view and shift what is considered important and what is valued. After all, Leopold referred to wolves as "vermin" before he saw the true results of aggressively pursuing his objective of a "hunters' paradise."

Is the concept of ecosystem services, in addition to serving as an aid to communication and in some situations as clarifying negotiations, capable of being a true transformative? Will it change the way people understand natural systems, and will it lead to new ways of valuing nature? Can it, for example, serve as the basis for an emerging and unified discipline of environmental evaluation? To these questions, also, I tentatively answer, "No." Ecosystem services should be understood as a tool, one useful in achieving goals of communication and collaboration but in most of its uses not a true transformative. It encourages single-loop learning; to achieve transformation—true double-loop learning—dominant metaphors must be challenged. While ecosystem services provide a novel way of communicating connections between human well-being and natural processes, the concept animates the same metaphor as the more standard economic view. It does not significantly alter the understanding of nature as a factory that produces raw materials for human benefit. Broadening the conception of human benefits is a very good thing; emphasizing the connection between ecologically based opportunities and the physical processes that support them is also a good thing—but as a useful tool, not a transformative force.

While the idea of ecological services enlarges our concept of human benefits, the concept still embodies the metaphor of nature as a productive system and thereby limits creative use of alternative metaphors. There is a legitimate fear that the ecosystem service concept, if used exclusively, will ultimately lock discourse into a narrow, monistic view of nature's value by further supporting the nature-as-productive-factory analogy.

8.2. Moral Imagination and the Role of Metaphor

What is needed is a broader discourse that digs more deeply than the rules of monistic systems to recognize that ethical decision making involves far more than applying rules to particular cases. Proceduralism has been a major theme throughout this book, but now it is possible to begin to understand how social values function and evolve within a broader, adaptive process of deciding what to do. What is needed is guidance from an emergent area of research in which cognitive studies overlap with ethics, seeking to learn how ethical thought and behavior actually occur in human organisms. An early result of this work indicates that the moral imagination functions to enable a decider to search through alternative scenarios, analogies, and metaphors and to determine the appropriate considerations "in a situation like this." This imaginative function is an essential part of moral decision making. Only after the moral imagination is engaged with a situation can the question "Which rule applies?" be asked.

This approach to understanding ethics more dynamically draws on important work on the role of *imagination* in reasoning about moral issues. This work emphasizing the imaginative function is more compatible with a pluralistic understanding, paying greater attention to the ways in which multiple values emerge and affect both public discourse and public choices in particular situations. This pluralistic, process approach need not deny a role for rules in moral reasoning; rather, it expands the moral realm to include an imaginative examination of alternative metaphors that, in turn, can guide the choice of an appropriate rule for a given situation. Criticism is still possible in this approach, yet the process of assertion and rejoinder is not viewed as algorithmic or syllogistic. It is not based on a single principle or formula but instead is seen as an intellectual negotiation that attempts consistency and partial consonance with multiple values and principles, relies more on sensitivity to empirical conditions shaping the context of the choice, and seeks balance instead of an objectively correct outcome. To quote Mark Johnson,

> Moral reasoning is a constructive imaginative activity that is based, not primarily on universal moral laws, but principally on metaphoric concepts at two basic levels: (1) Our most fundamental moral concepts (e.g., will, freedom, law, right, duty, well-being, action) are defined metaphorically, typically by multiple metaphoric mappings for a single concept. (2) The way we conceptualize a particular situation will depend on our use of systematic conceptual metaphors that make up the common understanding of members of our culture. In other words, the way we frame and categorize a given situation will determine how we reason about it, and how we frame it will depend on which metaphorical concepts we are using (Johnson, 1993, p. 2).

Traditional monistic systems of ethics see determining a moral outcome as a matter of applying to particular cases rules that were deductively derived from central principles. Our process approach, by contrast, places more emphasis on the empirical conditions of the situation, considers multiple possible analogies and metaphors with each one highlighting particular social values, and struggles toward agreements as to what to do in the face of multiple visions and characterizations of the problem. As these multiple visions and voices are expressed and brought to bear on the situation at hand, the focus is on seeking actions that can be supported by coalitions that find an acceptable balance in supporting competing values.

Monists, of course, will argue that such an outcome is too subjective and relativistic and will object that this approach leaves morality indeterminate and lacking in sharply delineated obligations. The answer to this charge is not to dispute it but rather to respond that human reason cannot and should not strive for syllogistic certainty. Monistic, objectivist approaches that claim to yield determinate answers to all moral quandaries overshoot what is possible in human communication and pursue a mythical goal of finding and enforcing a single correct response to every moral quandary; attempting to force moral reasoning into such strictures will miss most of what is important and interesting about moral reasoning. What *are* important are the stories that are told, the narratives often steeped in metaphor, that give meaning and context to individual actions and public policies.

According to this procedural approach that emphasizes the imaginative and narrative aspects of moral reasoning, metaphor "is one of the principal mechanisms of imaginative cognition. Therefore, we should expect our common moral understanding to be deeply metaphorical, too" (Johnson, 1993). As such, metaphorical reasoning rests "on imaginative structures that are *nonpropositional*" (Johnson, 1993, p. 102); such reasoning cannot be uniquely articulated in syllogistic form or involve inferences from one sentence or proposition to another. Leopold gathered an enormous amount of data, puzzled over it for several years, and then, in a virtual storm of creative thinking, reorganized both his data and his understandings by applying a new simile—"learning to think like a mountain." The key to understanding such a shift is to see that the new simile or metaphor adds new dimensions to the consideration. Leopold did not go from believing deer are an economic resource to believing they are not such; in fact, by adopting a new simile, he enriched his understanding of deer-wolf interactions with a broader perspective on the overall value of complex systems, including deer populations and their habitats. The simile of thinking like a mountain makes the situation more complex, adding a new layer of value and a new aspect of value for Leopold. As a hunter, he was forced to see that activity as conditioned by broader values that emerge at the level of the mountain (the ecosystem).

8.3. Ecology and Metaphor

To understand exactly what happened in Leopold's observation and reasoning in more detail, and with more explanatory context, I now turn to important recent work by the ecologist Steward Pickett and his colleagues.[2] Their work shows how metaphors play an important, if mostly hidden, role in ecology, especially in the application of ecological concepts in practical management contexts. They start from the assumption that all models are incomplete representations of reality and, accordingly, that there is no single, perfect model—all models focus attention on particular aspects of the complex systems that humans inhabit and study. While the term "model" is used in multiple ways in general discourse, Pickett and Cadenasso use it to refer to a representation of the systems and dynamics occurring in a given local situation (Pickett and Cadenasso, 2002). Further, Pickett and his coauthors have recognized that, as ecologists juggle and piece together partial models, their concerns and commitments affect the models employed. Therefore, any model used within a complex management context is partly a function of what is valued. Goals and concerns, as well as observable characteristics of the system being modeled, shape the chosen model. Metaphors are often the medium by which such models are built, explained, and studied.

These ecologists, then, have joined a number of others who see choosing appropriate metaphors for a given purpose as an important aspect of both scientific and normative thinking. Having thus ventured far past the boundaries of traditional positivist science, this group of ecologists espouses an aspect of ecology that is important to both theory and action—one that involves actively considering value-laden "choices" of guiding metaphors (Taylor, 2005). Pickett and colleagues have thus embarked on a most interesting experiment in postpositivist science.

Pickett and Cadenasso (2002) consider how one can choose a model appropriate to one's purpose, arguing that "[t]he richness of topics, complexity of model domains, and range of behaviors that models can exhibit suggest that ecosystem models can be used for diverse purposes" (Pickett and Cadenasso, 2002, p. 5; Pickett et al., 2004; Kolasa and Pickett, 2005). These diverse purposes include, of course, many descriptive tasks, which may well require diverse models, since models only partially represent reality. But these purposes also include another function of great, and undernoticed, importance: choices of models also reveal, and provide a conceptual entrée into, an imaginative realm in which scientists and practitioners seek appropriate metaphors to give basic shape and structure to a worldview capable of providing at least minimal levels of predictability. Predictability cannot be complete, so metaphors become a powerful tool for organizing data and gaining some level of control within a complex, dynamic, and uncertain world.

The reasoning of Pickett and his colleagues fits perfectly into the broader decision process framework being developing in this book. They argue that choice of model or models by which to understand management decisions is not syllogistic or algorithmic, but rather is a process in which multiple metaphors are proposed, hence the discussion—reminiscent of Simon's endorsement of procedural rationality—becomes one of choosing an appropriate metaphor that, in turn, will be useful in a decision process appropriate to the situation.

Pickett and Cadenasso (2002) write, "This area of communication includes education, the media, policy making, and management. In such public uses, the precision and narrow focus of technical terms is eschewed in favor of richness of connotation and in support of societally important, if sometimes controversial, values." They also advocate the identification of ecological systems with spatially defined areas and support the encouragement of recognition of systems as "places" with social meaning that endows residents with "responsibility and empowerment" (Pickett and Cadenasso, 2002, p. 6).

In describing the role of metaphors in the process of model-building, Pickett and colleagues emphasize their role in bringing ecological theory to bear in particular, problematic, and ecological contexts such as public concerns about the eutrophication of a lake or the clear-cutting of old growth forests. They explain that serious attempts to engage in ecosystem management must recognize three "dimensions" of the ecosystem concept: "meaning, model, and metaphor" (Pickett and Cadenasso, 2002, p. 2). The ecosystem concept consists, first of all, of an abstract, theoretical definition offered by Tansley (1935): an ecosystem is "a biotic community or assemblage and its associated physical environment in a specific place" (Pickett and Cadenasso, 2002, p. 2). This definition clearly places importance on local conditions, though it does not identify a particular place; likewise, this definition is without scale—it does not specify the boundaries of the system. In these senses, the definition of an ecosystem is too abstract to be of any use in managing actual ecological systems on a day-to-day or certainly a century-by-century basis.

For this definition to be useful in concrete management situations, it is necessary to build a "model," however crude, of how biotic and abiotic dynamics interact. Building of models is essential to bring the abstract definition of an ecosystem to bear on particular situations in which communities negotiate regarding the management of perceived problems.

This is the key point at which values and scientific facts interact in transformations such as Leopold's. The choice of models, which will determine the boundaries of the system brought under management, is affected (often invisibly) by one's concerns, values, and goals in understanding the system. Models, in other words, operationalize a certain perspective in a specific place, setting boundaries of the systems as described. These models can be essential in public discourse about what to do, as well as in framing problems and shaping possible solutions. So, the second dimension of the ecosystem concept emphasizes the choice involved when *a* model becomes *the* model by which a problem will be understood, discussed, and resolved. How are these important modeling choices made? They can determine the content of the scientific models adopted to understand real-world problems, and they seem to involve far more than descriptive aspects of the system: systems are modeled to highlight certain dynamics, because those dynamics are of interest to humans.

Focusing attention on this set of choices—those whereby scientists and others adopt a mental model of problematic biotic and abiotic systems—propels the argument to the third dimension of the ecosystem concept: metaphor. Those choices, if they are to avoid arbitrariness, must be considered imaginatively and as exemplifying varied metaphors. These metaphors guide the choice of appropriate models in highlighting dynamics that the model-builders associate with values embodied in key metaphors.

Pickett and his colleagues, as well as others who similarly recognize the importance of metaphors, have thus demonstrated that ecological understanding of particular systems rests on metaphors that connect changes in dynamic natural systems with human purposes and values, since metaphors inform models that allow both communication and understanding (Pickett and Cadenasso, 2006).

Another recent author has carried this line of reasoning forward by examining how metaphors in ecology interact with social conditions and values in dynamic settings. Brendon Larson, in *Metaphors of Environmental Sustainability* (2011), shows how metaphors that enter scientific, ecological discourse from ordinary thought and everyday speech can be intensified through "circular movement of feedback metaphors between biology and society." More and more scientists who have traditionally treated metaphor as rhetorical embellishments now see metaphorical thinking as central to the scientific enterprise. It is central, not simply in the static sense of being constitutive of accepted models, but also in allowing and encouraging dynamic changes in theories and hypotheses. Larson says, "Metaphoric ambiguity may actually be a fertile source of inquiry, and metaphors may allow leaps beyond stabilized meanings that revolutionize the process of scientific exploration" (Larson, 2011, p. 4).

Recognizing the inadequacy of the "linear" approach of developing science before addressing values and policy, Larson calls for a new form of scientific communication that treats the difference between scientists and laypersons as a matter of degree. They interact in a circular flow of conversation and of metaphors, from public discourse to science and back again (Larson, 2011, p. 17). Calling this function of metaphors "feedback," Larson recognizes that metaphors shape the way problems are framed, and these frames can trump individual facts. Through cyclical communication, however, even if metaphors are seldom explicitly chosen, it is possible to reconsider and modify these metaphorical understandings. It is here that metaphors provide the connection between science and public values, and in some cases each even intensifies the other in a positive feedback loop.

But Larson does not see only positive feedback—it is also possible for metaphors, when questioned, challenged, and modified, to drive forward the process of value articulation and value change. He writes, "There is a tendency for environmental metaphors to engineer a circular feedback between our view of ourselves and our view of nature. . . . Because of the authority of science in modern society, when scientists adopt a social term and give it new scientific meaning, they may simultaneously imbue its associated cultural values with new authenticity" (p. 18).

This dynamism can be the key to the criticism and evolution of public values: "The fundamental question, however, is whether these preexisting social values are consistent with sustainability. To the extent that they aren't, we need to reform our metaphors, even if we have to do so gradually. Without tremendous reflexivity and careful reflection, we merely strengthen the prevailing mind-set by implicitly endorsing it rather than questioning it" (Larson, 2011, p. 18).

Larson's activist and self-critical attitude toward metaphors that fail to highlight important public values can be seen as a mechanism for executing the critical systems theory (CST) of Ulrich, which challenges boundaries assumed in the models under use. Larson's discussion of the importance of making metaphors an explicit part of the public discourse about sustainability can exemplify Ulrich's goal of "emancipation" from partial and unproductive metaphors and the development of new metaphors and models that reveal values in a new light.

Larson's activist and evolutionary view of both science and society, taken with his belief that explicit reasoning by using metaphors can move a society toward sustainability, parallels an important piece of my reasoning in this book. My insistence that arriving at a sustainable future will require adaptive learning both in our beliefs and in our values creates a context in which Larson's understanding perfectly fits his methods. He provides one interesting example, the Gaia hypothesis: "The earth both is and is not like a living being, and we would benefit from living with the paradox of both of these views rather than gravitating to either extreme. The metaphor would then allow us to begin to query the dominance of the mechanistic paradigm" (Larson, 2011, p. 214).

Some readers may become cautious when Larson proposes a "code of conduct for the use of metaphor." He says, "Such a code would have to be flexible and seek an evolving middle ground between the extremes of objectification and catastrophism" (p. 219). One might wonder if this formalization of the process may be impossible or undesirable, yet Larson's belief that making metaphors more explicit, discussing both their scientific adequacy and their impact on social values, must be an essential aspect of any successful process that moves a society toward sustainable living. He states the "deeper lesson" of his book lies in "how metaphors can prompt dialogue between people with different perspectives. Sustainability will require the wisdom to be able to relate to one another" (p. 226). With this lesson, I can only agree, applauding.

Pickett and colleagues powerfully describe the essential role of metaphor in the development of ecology itself; Larson introduces a more activist and explicit process by which existing and proposed metaphors are to be vetted and criticized. While space does not allow a fuller development here of Larson's method, his approach does support the adaptive management process developed in this book.

Note, also, that this understanding of science and metaphor help explain the change Leopold underwent. Initially, under Pinchot's influence, Leopold was operating on the metaphor of nature as a productive system; since most of the land was too arid for vigorous forestry production, Leopold sought an alternative product that could be coaxed from the national forest lands of the southwest territories. The product he settled on was huntable deer, and the payoff would be in nature-based tourism exploiting the large deer herd, with improving success for hunters. This model, in turn, was based on a metaphor that treated deer as a product and thus analogized the ecological system to a productive industrial process. This analogy also, at first, blinded Leopold to the dangers of seeing ecological systems in this narrow, one-sided manner and in adopting the optimizing strategy without reservations. It also shaped the decision context such that the decision to maximize production of deer was addressed as a purely economic question having no moral implications.

While under the spell of his own original metaphorical model of nature, Leopold—as would be expected, given the above analysis of the role of metaphor in moral reasoning—never questioned that he should apply the utilitarian rule in his dealing with wolves, mountain lions, and deer. So he adopted what turned out to be a disastrous course of action that eventually undermined his own goal: "The starved bones of the hoped-for deer herd, dead of its own too much, bleach with the bones of the dead sage, and molder under the high-lined junipers" (Leopold, 1949, p. 132).

An economic maximization rule dominated his thinking because of the shaping power of the underlying, dominant metaphor of nature as productive system. Indeed, this metaphor so shaped his thinking that a "conversion experience"—an encounter with a dying wolf—was necessary so that he could begin to see natural systems in a new way. Disappointment in the performance of the artificially productive system (deer on wolfless ranges) forced Leopold to consider alternative metaphors. Provoked by a sense that dying wolves were trying to tell him something, he articulated his brilliant simile, "thinking like a mountain." He had to break out of the dominant, ecosystem-as-productive-system metaphor to see the flaw in his practices. Doing so also involved the awakening of his long-term concern for the stability, health, and resilience of the system.

Like Pickett and his colleagues, Leopold did not merely challenge details of his economic analysis, he also challenged the dominant metaphor itself. On declaring that we must learn to think like a mountain, Leopold introduced a new barrier-busting simile. By replacing the productive system metaphor with the metaphor of a multilayered, complex system, he countered the economic metaphor with a more complex ecological metaphor in the form of an expansive simile.

Leopold, then, anticipated the insight of Pickett and colleagues that metaphors, analogies, and similes from ecology can play a crucial role in determining the system studied and the values highlighted. He also anticipated the expanded role for ecologists sketched by Pickett and coauthors: ecology can contribute new and more appropriate metaphors, and it can adjust the boundaries of systems. It can thus play a crucial role in revealing new and more appropriate values and ways to act on those values.

Leopold's simile of thinking like a mountain does not signal a simple shift from one dominant conceptualization of environmental issues to another. His change, on the contrary, caused him to see the resource situation in more complex terms. He did not stop hunting or encouraging outdoor tourism;

rather, he considered hunting success and the well-being of the "mountain" as alternative values, both important. Deer are not either an important economic resource in the American Southwest or a threat to ecosystems and wolfless habitats. Deer are both. They play an important role in multiple narratives, and a full understanding of the decision situation faced in the Southwest regarding deer herds requires that deer be seen as both—a perceptual challenge that is supported by multiple metaphors and multiple stories, as illustrated in figure 4.3.

If one tries to understand Leopold from a monistic viewpoint, assuming one right solution to problems based on a central principle, one will see him as facing a dichotomous choice between hunting and protecting the mountain. But if one understands Leopold pluralistically, as recognizing alternative metaphors both of which have some validity but none of which can tell the whole story of mammal management in the Southwest, one can see that Leopold saw the task as balancing multiple values. On this new viewpoint, Leopold could embrace these multiple stories as each contributing to the dialogue and to the consideration of what to do. These stories and associated metaphors contribute to a process, one in which ecological data and ecological theory broaden and deepen the discourse, allowing a community to engage in discussion, sharing of views, and—one hopes—social learning about the complexities and the possibilities of living within complex, multilayered ecological systems.

8.4. Can the Many Metaphors of Conservation and Ecology Be Integrated?

The metaphors that underlie and support one's worldview determine whether one looks at environmental problems expecting to see an impending tragedy or to view a mountain with many levels and layers of life. Neither the orientation that sees market-driven tragedies nor Leopold's holistic vision of a multiscaled, complex mountainside, tangled with interconnections of both causation and value, is objective reality. These are two metaphor-driven viewpoints and, while there are crucial data-points—fluctuations in deer populations, for example—it was a metaphor that led Leopold to focus on an entirely new set of issues.

Moral monists, who intend to emulate Leopold, insist that he first adopted an exclusively economic value measure and then switched to the opposed position: that nature has intrinsic value that trumps and invalidates human instrumental uses. This simplification misses the crucial point: he saw the challenge of economic development in a semiarid region as a problem of integrating these values rather than choosing between them. On his altered perspective, both deer and wolves have a recognized value, but those values must be understood as occurring on a different scale in space and time. Managing a deer herd to increase hunting-related tourism requires a time horizon of two to three years: "A buck pulled down by wolves can be replaced in two or three years." Recognizing the value of wolves requires a time-horizon of ecological time, however, because "a range pulled down by too many deer may fail of replacement in as many decades" (Leopold, 1949, p. 132). Leopold transcended competition between these two management regimes as he sought policies that would integrate both managerial challenges by seeking a multiscaled policy that will protect both values. In practice, after 1935, Leopold consistently stated his policy of maintaining healthy wolf populations in wilderness and other wild areas with limited human population, though he never advocated full reversal of wolf removals.

What he did advocate (without ultimate success in his lifetime) was a multiscaled policy that would protect wolves in wild areas, so that he could meet the challenge of maintaining a critical piece of diversity, meanwhile hoping the wolves would at least control the deer populations too remote to be heavily hunted. This was a key aspect of the management problem, since remote populations of deer that were freed of predation radiated outward from lightly hunted ranges and exacerbated the overpopulation of deer throughout the region. On the other hand, Leopold never rejected hunting or managing deer populations for local hunters and tourists. Values that, on a monistic viewpoint, seemed diametrically opposed each found their place in a multilevel, integrated policy.

This policy incorporates two basic metaphors together by seeing them as aspects of the system that function at different scales and are driven by different dynamics. While Leopold's insight was fully worked out long after leaving his managerial leadership in the Southwest, his accomplishment was to show how a multiscaled understanding of ecological systems can bypass some of the worst dilemmas of management, provided that the data, the questionable theories, and the governing metaphors are all adjusted. Leopold illustrated how to conceptualize more than one metaphorical understanding in an overarching, multiscaled system by using the thinking-like-a-mountain simile to integrate multiple perspectives, each of which incorporates processes that serve multiple human interests via processes taking place at different scales. If one manages things in this way, the old conflicts between economic development and environmental protection can be finessed, since an integrated system protects both. The result is a way of thinking about value such that one must keep score on multiple levels, seeking policies that will find a reasonable balance among competing uses of ecosystems. Adapt's tenth heuristic urges pluralism of metaphors.

Heuristic 10

The fitting many metaphors heuristic. Avoid single, master metaphors. Metaphors close off as well as open up vistas and perspectives; therefore, work with a shifting kaleidoscope of metaphors that will also track shifts in language and conceptualizations. Seek integrative metaphors (e.g., Leopold's thinking like a mountain) that are inclusive of more visions and encourage a balance among varied social values flexible enough to evolve as situations change.

Having criticized monistic theories as necessarily failing of completeness—a major theme of the first half of this book—and having cast our lot with pluralism, we now can see that, when facing a problem that manifests within a complex, multilayered system, pluralism of scales can actually open up opportunities for an integrative perspective that associates differing values with differing dynamics. Further, it is possible to see the advantages of pluralism by recognizing the opportunities for both value dynamism and the integration of values associated with different ecological dynamics that unfold at different scales. As Pickett and colleagues show, it is through metaphor that the scope and scale of problems is determined, and it is through metaphor that a descriptive, ecological understanding of an on-the-ground problem is mapped onto social values so as to support an integrative perspective.

In fact, of course, the history of conservation biology and conservation ecology is replete with multiple metaphors and analogies. John Muir and other nature writers and advocates, just like Aldo Leopold, often explained their points, and tried to transform their colleagues' understanding of situations, by invoking metaphors. For the Pilgrims, nature was foreboding, satanic, and symbolic of evil; for Muir, forests were "cathedrals." Others, on their way to a stewardship ethic, see nature as a "garden."

Leopold's writing, in particular, features a flood of metaphors and analogies used in various circumstances to make specific points. Learning to think like a mountain was the metaphor he used to integrate different theories and models that he developed to apply at different scales. He accepted the exploitation of a resource such as deer, but only provided that those activities do not disturb the functioning of the larger mountain (symbolizing the encompassing ecological system). When concentrating on what should guide the activities of managers and users of land, Leopold employed medical analogies—health and integrity—to encourage land users to ask serious questions about how their activities affect the ongoing functions of systems.

As Julianne Newton (now Julianne Warren) has effectively argued, Leopold used many metaphors, but in the end he relied on the central metaphor of land health: "Leopold's guiding land ethic and the goal of land health could be applied to the use and conditions of all lands. On the other hand, no two land parcels were alike. . . . On some lands the organization of the land pyramid appeared to be more flexible and able to adjust to change without becoming deranged. A particular action on one landscape might be right while the same action elsewhere might be wrong" (Newton, 2006, p. 349).

While elsewhere in this passage multiple metaphors are mentioned— "land health," "land pyramid," "become deranged"—Newton identifies "health" as the guiding metaphor: "In the end Leopold was able to address with his land ethic nearly the full range of tasks that had drawn his attention. It pointed toward the overall goal of conservation—land health—offering the prospect of redressing the fragmentation that so afflicted the conservation cause" (Newton, 2006, p. 349). Use of health and integrity metaphors later in his career further document Leopold's developing understanding: the health of an organism implicitly refers to multiscalar organization, since functioning organisms must be healthy at each scale, from individual cells to the whole organism.

These many metaphors and comparisons, however, have not added up to a dominant, master metaphor that yields the singular structure of nature and natural systems. This is what should be expected, given the incompleteness of any one model for reality. Sandra Mitchell, a philosopher of science who examines the special features of complex systems, has been driven by apparently incontrovertible arguments toward a pluralistic understanding of those systems: "pluralism in causes, pluralism in ontological levels, and pluralism in integration will characterize the scientific explanation of complex system behavior." Further, she argues that, while these pluralisms must ultimately be "integrated," the relationships cannot be resolved "algorithmically" (Mitchell, 2009, p. 110). She states that pluralisms come in multiple versions. First, she mentions and rejects what she calls "anything goes" pluralism, which she attributes to Paul Feyerabend, arguing that this pluralism goes too far toward "nihilism about effective investigatory methodologies." The epistemology for adaptive management is here seen to have a strong enough epistemology to avoid this relativistic version of pluralism, so it can be concluded that the commonly cited argument against pluralism (that it is relativistic) can be rejected. This form of pluralism would not do for the purposes of adaptive management. Second, she discusses Philip Kitcher's version of pluralism, which she calls "winner-take-all pluralism," which sees pluralism as a temporary status in the development of knowledge since he expects scientific progress to eventually eliminate all but one of these. She describes this version as "a pluralist methodology for achieving a monist goal." Third, she cites "partial" integration, which is achieved when differing theories explain different aspects of a complex system, in ways that involve both competition and complementarity of the theories (Mitchell, 2009, p. 107–8).

Yet these forms of pluralism are not what Mitchell is going for; instead, she argues that detailed understanding of complex systems and of scientific study of them suggests a richer kind of pluralistic integration. This form of pluralism "attempts to do justice to the multilevel, multi-component, evolved character of complex systems" (Mitchell, 2009, p. 114; also see Norton, 2005). This richness, which stems from the contextual nature of relationships in a multileveled system, prohibits generalization of the details that are operative in specific situations. This contextuality is total: the way that a diversity of factors are integrated with emerging models of system behavior is itself context-sensitive. The reconciliation achieved in specific situations, like Leopold's simile of a complex mountain with multiple levels hierarchically ordered, is effective if it works to bring, if not closure, then at least insight with respect to the system in question.

Skeptics may jump on this point. If there are multiple explanatory frameworks that might reconcile multiple levels and layers of explanation, and if choices to study these depend on the context, a critical skeptic will argue that any resolution of apparent conflicts and integration of multiple levels will be "relative to context." Can this form of the relativistic criticism be answered?

Yes, it can. Suppose a researcher with no interests or values in some context, with zero investment in the system at hand to be studied, tries to develop an integration of the multiple complex processes occurring. This researcher will indeed be without basis for achieving the kind of enlightening reconciliation that Leopold enjoyed, so the researcher might conclude that "anything goes" in characterizing the system. There is no correct way to put parts of complex systems together in models unless a perspective is adopted, such as that of an adaptive manager who is tasked with managing deer populations. The key ingredient in Leopold's transformation—and, by hypothesis, of other such integrative transformations—was a commitment to learn about and manage deer populations. This gave him his original orientation toward the ranges of the Southwest as a productive factory. When the actions he based on this metaphor led to starving deer, his orientation changed and he adopted a new orientation to the evolving problem by choosing a metaphor appropriate to his evolving managerial goals and attitudes.

I have argued in 2005, and in the present book, that choices of perspective and orientation necessarily involve what can be called "pragmatic" choices. Leopold's complex model of the mountain was not *the* correct model, any more than his model of ranges as production facilities for deer was correct. Leopold's thinking-like-a-mountain metaphor became appropriate to the context when he found that acting on the productive factory model was inappropriate. The latter included only a single scale of the system dynamic, a simplicity that thwarted Leopold's plan for a "hunters' paradise."

Appropriateness in these contexts depends on the actor's purposes. In philosophy, these choices of how to describe and model systems—of how to create tools by which to understand a complex system, given a human purpose—are called "pragmatic" choices. Leopold's choice of the mountain metaphor was, in keeping with procedural rationality, "appropriate," since he was managing the deer herd for hunters and he wanted the system supporting the deer to remain productive indefinitely. In view of Leopold's commitment to manage to protect economic, ecological, and evolutionary processes, his simile and the associated holistic metaphor are both perfectly reasonable, even though that reasonableness depends on his personal values and goals.

We can now generalize. Ecological systems and systems that mix human and natural elements are multileveled as well as complex. This means there can be no single, correct representation of the systems, nor can there be a complete master metaphor that reveals an objective reality behind our many partial representations. A procedural approach, however, which seeks reasonable and appropriate methods, and combines this sense of appropriateness with a recognition of the importance of the goals and commitments that focus researchers on a given level, can encourage them to seek out the right scale and the right granularity for addressing their concerns. Specific contingent characteristics of a situation set the context, and one can only make judgments about what is "appropriate" by paying attention to the pragmatic concerns of inquirers.

This point is summarized by Mitchell as follows:

When should we use the fine-grain representations and when could we use coarser-grain representations? They are both "true" and "accurate" ways to capture features of the world we want to represent. This is the crucial point:

what determines the level of abstraction or granularity is pragmatic *and can be answered only with reference to the particular context and scientific objective*. Different representations will be better for solving different problems. What we should expect in science is a diversity of representations, in addition to the plurality of hypotheses attaching to different causes and different levels of structure in complex systems. (Mitchell, 2009, p. 117)

CHAPTER 9

Adaptive Collaborative Management: Empirical Findings and Case Studies

9.1. Pluralism and the Multigenerational Public Interest

I set out to examine, from a fresh perspective, the connection between understanding sustainability and evaluating environmental change. This perspective eschews upfront arguments about the nature of environmental values in favor of examining the nature of environmental decisions and, especially, the actual and possible processes by which communities struggle, or could potentially struggle, to articulate and pursue more sustainable policies. My approach is based on a well-established empirical generalization: in most situations, especially ones in which there is spirited disagreement about what to do, participants in the discourse espouse and defend a variety of values (Minteer and Manning, 1999). Disciplinary theorists, encountering this pluralistic situation, see it as problematic, so they set out to develop "reductions" of other values to the kinds of values they favor theoretically. For example, when economists encounter a participant who says, "I support the protection of the Alaskan National Wildlife Reserve and oppose oil exploration there because the ecosystem and the caribou herds have intrinsic value," they try to determine how much that person (who does not even live in Alaska) is willing to pay for that protection, registering this value as willingness-to-pay (WTP) for an "existence value" (the value one places on simply knowing something exists). By thus representing the value of caribou protection as aggregated individual WTP, they succeed in reducing that value to an economic value and in the meantime enrage nonanthropocentrists for turning nature into a collection of commodities.

My approach has been to replace these intransigent theoretical and ideological debates within a deliberative process and to continue our examination by paying attention to the processes by which affected parties in a dispute interact. By focusing on interactions within a process, attention is directed at decision making more than at motivations of participants, and at deeds more than words. Here, I have helped myself to a working hypothesis: that better processes will lead to more cooperation and better decisions. This hypothesis encourages pluralism and focuses more on the processes by which the individuals and groups, serving differing and sometimes opposed values, move toward action. Hence, I have emphasized the nature of problems and the tools available to rationally address complex problems.

Value pluralism regarding the environment is thus a political problem—one of how to make better decisions, even in situations where individuals, groups, and interested parties do not share a common viewpoint on the values of nature. The problem of finding a route to sustainable living thus becomes a matter of deliberation and negotiation. While accepting pluralism, these deliberations must be normative: the goal of environmental public policy, after all, is to serve the (procedurally defined) public interest. Learning to live sustainably, then, is learning how to project the public interest into the future for multiple generations. It requires a concept of the public interest for a Burkean society: the eighteenth century philosopher Edmund Burke defined a "society" as "a partnership not only between those who are living, but between those who are living, those who are dead and those to be born" (Burke, 1910, pp. 93–94). So the search is not pointless: one can anchor these processes and associated efforts in the idea of a multigenerational community guided by procedural rationality toward a multigenerational public interest.

This idea of a multigenerational public interest, by the nature of the public interest as introduced in chapter 5, will not identify a single measure or a single variable that can be projected into the future. On the contrary, the Deweyan conception of a public interest involves an ongoing process of deliberation and negotiation, as pictured in figure 9.1. Sustainability so understood must emerge from iteration and be constantly altered and updated by it. This chapter explores some of the empirical literature evaluating the efficacy of collaborative processes and introduces case studies that illustrate how these processes can operate in real situations.

A rapidly growing literature continues to examine and evaluate ongoing collaborative processes, though this area of empirical study is challenging. It would be ideal if social scientists could provide evidence that some specific processes have usually led to better outcomes (understood as more

Figure 9.1. Strategies of Optim and Adapt: Optim pursues a path in search of an optimal objective, while Adapt considers more possibilities and experiments to determine which one is best.

effective protection for natural systems) than would be expected without the process. This direct approach to empirical study, however, faces several problems. First, many processes begin before any baseline data is gathered. When scientific researchers join a process, actions are already under way, and they cannot compare outcomes with a baseline. Second, each problem and responding process exists within a unique context; this means there will never be perfect and seldom even partial controls. Social scientists can make judgments that a situation is getting better or worse over time, but they generally do not have sufficiently parallel control cases to allow evaluating some processes as superior to others in realizing environmentally protective outcomes. Third, most environmental problems have emerged over many years, so research projects are almost always too short in duration to observe definitive results about outcomes. In sum, scientific attempts to compare ecosystems, either with other ecosystems or across time, suffer from difficult problems stemming from the problems just mentioned. Often, researchers simply cannot stay within the usual constraints of the scientific method, and yet, using the best information available, they can nevertheless learn a good deal about systems by observation and comparison. Often, their efforts are best described as "quasi-experiments."

As a result of these empirical challenges, studies of collaborative processes have generally taken the form of case studies of single processes or multiple, comparative case studies of a selection of processes, using mainly interview and questionnaire methods. Lacking controls and operating within the duration of funded research projects, researchers have found difficulty in judging the success or failure of these multiproject studies. Researchers, by relying on the memories of interviewed participants, can obtain information about the beginnings of processes and events that took place before the research project. The information gathered in this way, however useful, generally does not allow the researcher to conclude that key outcomes occurred *because* of the process, due to the absence of baseline data and parallel control situations. Additionally, it is difficult to eliminate the bias individuals have toward processes in which they are invested, so participants' self-reports can introduce bias into empirical studies.

Despite these problems and limitations, I proceed in this chapter to examine the existing case-study literature. I begin with a brief exploration of the motivations that can lead groups to undertake the harrowing task of forming collaborative organizations, setting rules for individual behavior, and developing institutions to enforce those rules. These preliminaries will set the stage for a brief summary of the empirical literature on adaptive collaborative processes in section 9.2. In sections 9.3–9.5, I discuss two detailed case studies of processes for which we have long-term data and information.

9.2. An Overview of the Literature on Evaluating Collaborative Processes

Since making incrementally better decisions, as we have learned, is essential to democratically developing sustainable goals and policies in particular societies and communities, empirical testing of the effects of various types of processes holds one key to learning our way to sustainability. One important role of adaptive management is to identify what works. As it turns out, there are many group processes operating in many locales and regions of the United States, and the burgeoning study of these group processes by social scientists provides a rich source of data and insights about what works and what does not work as groups form to address environmental problems. I highlight a few of these cases in a brief, unsystematic survey of the empirical literature on the value and fate of adaptive collaborative management processes.[1]

I intend to discuss these processes inclusively, observing a variety of arrangements that involve "committees, councils, advisory groups and task forces" (Leach et al., 2002, p. 1). Many collaborative processes are watershed partnerships that exist on the US West Coast in great numbers (Leach et al., 2002). Because watersheds are hierarchically structured, these groups might include watchdog groups that exist at different, embedded scales.

What criteria can be used to evaluate specific processes? Every community faces a distinct set of circumstances, so it is difficult to establish controls that would be necessary to establish causality between particular aspects of processes and outcomes. This observation is pointedly made by Judith Layzer (2008).

After summarizing important advantages of participatory processes involving stakeholders—which seem to increase human and social capital in a community and lead to more agreement among the stakeholders—Layzer cautions, "Researchers have not discerned a clear relationship between these two central achievements, however; in fact, some analysts suggest that funding levels and passage of time, rather than trust and social capital, are the keys to successful implementation" (Layzer, 2008, pp. 2–3).

The difficulty of drawing general conclusions regarding causal relationships is inherent in the case study approach and is apparently unavoidable, given that cases vary in so many ways, making closely parallel controls impossible. In particular, no researchers have successfully measured a relationship between specific sets of processes and specific outcomes observed as measurable improvements (or decrements) in the quality of an environment.

Nevertheless, researchers have developed a variety of techniques and criteria for studying the impacts of management processes. Results vary widely, both because researchers define the categories of management processes somewhat differently, resulting in different comparisons being made, and because of many local variations on the theme of cooperation. For example, researchers studying a case in Nevada County in California found that powerful interest groups set out to "contain" and to "capture" the cooperative processes to achieve goals outside and contrary to the objectives of the process. When these political forces failed to control the process, they derailed it, resulting in a "war" between environmental and development interests (Walker and Hurley, 2004). Keough and Blahna (2006), however, surveyed a wider range of cases presenting four detailed case studies and concluded that, in general, more collaborative, adaptive approaches improved local resource management decision processes. They identified eight characteristics that improve decision processes: (1) integrated and balanced goals, (2) inclusive public involvement, (3) stakeholder influence, (4) consensus group approach, (5) collaborative stewardship, (6) monitoring and adaptive management, (7) multidisciplinary data, and (8) economic incentives. They then showed, using their four case studies, how these features encourage a better balance among social, economic, and ecological objectives.

From a survey of the many case studies in the literature, it is possible to tell stories about cases in which cooperation, collaboration, participation, and adaptive use of science have apparently led to better decisions; it is also possible to relate stories in which powerful interest groups have formulated questions in ways that prejudice the case toward development, stymied cooperative actions, and insisted on lowest-common-denominator environmental protections. There is no magic formula that, if followed, will always result in improved processes and in improved decisions, nor is it true that in every case powerful private interests will be able to thwart the public interest. This is what we should expect in the pursuit of procedural rationality: that if processes are appropriate, one can expect positive results. But if inappropriate processes are pursued—perhaps because bad processes predominate or because some members of a partnership engage in bad-faith attempts to subvert the process—outcomes can be expected to disappoint.

Given these mixed results, social science research should focus even more intently on identifying those aspects of management processes that encourage, and those that weaken, participatory, collaborative processes so as to bring the power of empirical social scientific research to bear on the problem of creating more effective processes. By learning how to make better decisions, especially those with long-term consequences, one can learn how to learn to be more sustainable.

To this end, I turn to a more detailed examination of two systematic bodies of research on the impacts of collaboration on policy processes. One group of researchers, led by Paul Sabatier at the University of California–Davis, has undertaken the most extensive study of cooperative management processes, concentrating its efforts on the many watershed partnerships that have emerged in California and Washington state. The prevalence of such processes is testimony to their popularity. Sabatier, Leach, and colleagues refer to a study that estimated that there are 346 such partnerships ("probably an undercount") west of the Mississippi River, from which they chose their original forty-four cases randomly (Leach et al., 2002). A second study, a systematic, single-authored study comparing seven case studies by Judith Layzer, provides a somewhat different perspective.

Sabatier and associates have, in their work, articulated and applied six criteria to evaluate the performance of partnerships formed to improve management of watersheds. They used interviews, surveys, and documents used in the process to measure the following:

1. Perceived impacts of the partnership on specific problems in the watershed
2. Perceived impacts of the partnership on human and social capital
3. The extent of agreement reached among the stakeholders
4. Implementation of restoration projects
5. Project monitoring
6. Education and outreach projects (Leach et al., 2002, p. 2)

It has proven very difficult to devise evaluative tests that causally relate aspects of a process to environmental outcomes. Sabatier and his coresearchers, including William Leach, have responded to this difficulty by using semistructured interviews and questionnaires to determine whether participants themselves self-report (1) positive impacts of their partnership on specific problems and (2) positive impacts on human and social capital. These researchers are well aware of the difficulties of validity that may arise in basing conclusions about successes and failures of partnerships on the self-reports of participants. They advance their positive assessments of outcomes with regard to the above criteria 1 and 2 tentatively, mentioning that these results may be partially generated as a result of reverse cognitive dissonance, sometimes called the "halo effect": participants who invest time and effort in a process are unlikely to describe the process as worthless or harmful (Leach et al., 2002, p. 17; Layzer, 2008, p. 296).

Other criteria can be tested more concretely, as it should be possible to get reliable evidence about the degree of agreement achieved among stakeholders, the implementation of restoration projects, project monitoring, and education and outreach programs undertaken. Problems with perception-based data could be overcome if baseline data was collected early and if means of testing hypotheses regarding changes to key variables were projected. Sabatier and his group strongly recommend that in the future groups gather baseline data on the condition of the watershed at the outset and undertake periodic self-assessments, but it is often the case that startup organizations lack funding to undertake monitoring immediately. Until this kind of studies become more prevalent, however, "researchers interested in comparing success across partnerships will need to rely upon both objective measures of activities and perceptual measures of impacts" (Leach et al., 2002, p. 18).

Based on their assessment of an expanding, randomly chosen set of actual watershed partnerships, this research group states the following conclusions (while offering the above caveats about the validity of perceived effects). First, they found that partnerships have positive impacts on what are perceived to be the most serious problems affecting the watershed in question. This evidence seems to contradict the claim, made by some critics of collaborative management efforts (Layzer, 2008), that collaborating groups tend to avoid difficult, controversial problems.

They also note that time of duration has a strong (positive) effect on the perceived and measurable impacts of a process; so it important to compare partnerships of roughly the same age. While this age effect may be partly a result of the halo effect, there is evidence that longer-lasting partnerships do become more active and effective over time. Criteria 3 to 6 provide important measurable data showing that, over time, watershed partnerships undertake more projects and become more effective in shaping the society's response to environmental problems than occurs in their absence. Leach and colleagues say that it is important for funding agencies and public officials not to judge early-stage partnerships as failures too soon: "It takes time (typically 4 to 6 years) to educate participants, overcome distrust, reach agreements, secure funding, and begin implementation" (Leach et al., 2002, p. 17). To the extent that one can treat cooperative actions to improve the local environment as evidence of success of groups, observations and records of actions such as monitoring and education programs provide nonsubjective evidence of success.

The Sabatier team has so far studied dozens of watershed partnerships. Their research has shown that there are far more of these processes, of varied degrees of formality and flexibility, than most observers have assumed. They also cite evidence that ongoing processes result in improved public understanding of environmental problems, development of significant social capital supporting efforts to protect the environment, and growing understanding that environmental problems must be addressed more holistically. Leach et al. (2002), found strong evidence that participants believe strongly in the collaborative processes they engage in, as 84 percent of participants "agree or strongly agree" that "the best strategies for resolving watershed issues involve consensus-based processes" (p. 6).

While evidence of actual environmental improvements is difficult to assess, these researchers have clearly established that, at the very least, participatory efforts can contribute mightily to the dynamic aspect of environmental evaluation, as introduced and emphasized in chapter 6. Thus there is strong data to support the view that opportunities to meet, discuss, and interact with scientists and stakeholders from different perspectives can transform public perceptions. By encouraging a more holistic and scale-sensitive understanding of environmental problems, such deliberations contribute to the ongoing process of incremental learning so essential to adaptive management, as well as to the dynamic, transformative effect of scientifically based adaptive management. So, while there remains considerable controversy regarding whether collaboration and flexible, adaptive management can be empirically demonstrated to improve on-the-ground environmental management as measured in improved outcomes, there can be no doubt that these processes contribute to social learning on their way to sustainable living.

The planning scholar Judith Layzer (2008) has attempted a more critical analysis of the effects of stakeholder collaboration by focusing on new approaches to landscape-scale governance. To avoid the problems of halo bias associated with asking participants to assess the effectiveness of the processes in which they are involved, Layzer introduces an alternative methodology for assessing environmental outcomes. She exploits the fact that most formalized collaborative processes involve consultation with scientists who form a "technical committee" to assess problems and threats as well as to offer possible remedies. Layzer then evaluates plans and outcomes generated by processes according to the likelihood of success in achieving goals (such as protecting biological diversity or expanding a key type of habitat) proposed by the group's technical committee. For example, if the technical committee recommends protection for an area of N acres, with a buffering system around the core habitat areas, but the actual plan halves the acreage set-asides and makes no provision for a buffer to protect the core habitat area, the plan will be judged—on the basis of the technical committee's expert opinion—to be unlikely to provide adequate protection. If, on the other hand, the plan resulting from the collaboration fulfills all or most of the technical committee recommendations, Layzer would judge this to be an environmentally more successful process. She has thus introduced a measure of success that displaces the subjective element of the perception of participants with a set of standards that are supported by experts. This is not as reliable a method as would result from beginning a process with baseline studies followed by periodic assessments of outcomes, but it does include data derived from a source other than impressions of individual participants.

Armed with this methodology, Layzer claims the following:

1. Processes with more stakeholder involvement result in greater social capital (here agreeing with the findings of Sabatier's group).
2. Such processes also lead to greater agreement among stakeholders about science.

But she also has a third claim:

3. Stakeholder involvement in setting goals of management systems often leads to less-stringent policies to protect nature, as the desire to find win–win solutions that appeal to everyone waters down any measures strong enough to actually protect the resource involved.

Note, first of all, that from the viewpoint taken in this book—an emphasis on the dynamic, transformational aspects of the evaluation process—the positive findings regarding claims 1 and 2 are extremely important. If the goal is a transformation, through social learning of the public's tastes and beliefs—a transformation that will create an improving conception of sustainability in the population—then success on Layzer's first two claims might be of paramount importance.

At first glance, however, Layzer's third claim—leaving aside the gains in social capital and social learning—may seem to undermine significant arguments of this book. For example, she argues that involving all stakeholders from the outset is likely to lead to watered-down environmental goals, as the statement of objectives will include no objectives that impose costs on any interest group. She also argues that introduction of new institutions to supplement existing jurisdictions leads to serious governance problems. Finally, she argues that, using her expert-based methodology, collaborative decisions tend to underprotect important resources by choosing a plan that imposes minimal costs on affected parties.

A careful analysis of Layzer's argument, however, reveals a much more complex relationship to the argument advanced in this book. Without challenging the excellent data she provides, I must reanalyze that data if it is to be applied to our argument. Using the case study method, she divides her seven case studies into two groups, one of which she refers to as "ecosystem-based management" (EBM) and another model that she calls "grassroots ecosystem management" (GREM). The four cases she calls EBM are very large-scale, regional management efforts that have usually been initiated by the US federal government and involve creation of new institutions (such as a steering committee) created independent of existing institutions and jurisdictions to manage the process and relationships among the involved groups.

Layzer defines the differences between the two formats she wants to study as follows. GREM approaches, she writes, "aim to change a culture rather than the rules of a place, are typically initiated by residents of rural, western communities threatened by disputes over natural resource extraction, mostly on public lands. By contrast, EBM—as more commonly defined—tends to be instigated by government officials and seeks to institutionalize new forms of governance to address pollution and resource management problems. EBM initiatives span large landscapes that may encompass marine or other aquatic ecosystems, publicly and privately owned land, and urban as well as rural areas" (Layzer, 2008, p. 2).

Layzer's EBM case studies consisted of two terrestrial system plans— Austin, Texas's Balcones Canyonlands Conservation Plan (BCCP) and San Diego's Multiple Species Conservation Plan (MSCP). The case studies also included two large-scale aquatic projects—the Comprehensive Everglades Restoration Project (CERP) and the California Bay-Delta program (CALFED). For contrast cases, Layzer chose Pima County, Arizona's Sonoran Desert Conservation Plan (SDCP); the Kissimmee River Restoration Project (Florida); and the Mono Basin Restoration (California). Based on her research on these seven cases (four EBM cases and three cases that better fit the GREM model), Layzer draws several conclusions that seem relevant—both positively and negatively—to my discussion of adaptive, collaborative management in this book.

Speaking of positive impacts of adaptive, collaborative management, she concludes the following:

1. A landscape-scale perspective improves the likelihood of
   a. increased cooperation among agencies that have in the past worked at cross-purposes (p. 270), and
   b. developing an integrative scientific assessment and creating a more comprehensive plan (p. 270).
2. Stakeholder involvement has positive effects, because
   a. deliberation in search of consensus can transform perceptions of individual interests, creating trust among participants and leading to more openness to innovation;
   b. learning through science replaces bickering about untested scientific hypotheses; and
   c. participation in developing a plan increases likelihood of implementation because the participants gain a sense of "ownership" of the plan.

Interestingly, however, she ends with a strong critique of EBM, arguing that three features of the approach lead to failure to protect biological diversity and important habitats:

1. Landscape-level planning, when combined as it is in EBM with the creation of new institutions in which stakeholders are included and decision making is based on consensus, allows politically powerful interest groups to redefine the problem away from protection of species and habitats, transforming the question into one of how to continue rapid development without breaking federal laws.
2. While landscape-level planning initially led to better coordination, this cooperation was attenuated over time as some interest groups come to believe they can achieve better self-interested results by "going it alone."
3. The new institutions and organizations developed to allow large-scale management decisions affecting whole landscapes, being based on consensus and group formulation of the questions faced, are not as effective as conventional political institutions in developing plans with real teeth; local political activism, local leadership, and the use of litigation are thus favored by Layzer over more consensus approaches, especially at the scale of EBMs.

Before endorsing Layzer's list of advantages and criticisms, however, several clarifications are in order. First, her target of EBM, and also her inclusion of the three comparison cases, implies a narrower focus than I have in this book. In fact, I think she can only be correctly understood by recognizing that all seven of her landscape-scale cases, whether best described as EBM or as GREM, fall within the purview of adaptive, collaborative management as understood in this book. I, unlike Layzer, am arguing for a broad trend in environmental management that departs from the standard understanding of scientific resource management (SRM) that began with Gifford Pinchot late in the nineteenth century and ended with numerous public objections and court cases as new legislation created a new environmental regime in the 1960s and 1970s. Layzer's seven cases, though strongly contrasted in her book, are all included in the much more general trend that is being explored here. This explains how she can recognize important advantages of collaboration, holism, and flexible adaptive management in general terms—she is here recognizing the unquestioned improvements attendant on the larger trend that has changed policy away from atomism and toward integration since the 1960s. Accordingly, I can endorse the list of advantages Layzer provides and accept her empirical data as at least partial support for my advocacy of the adaptive collaborative management approach over the discredited SRM approach.

Layzer's list of advantages of what she refers to as the "landscape-scale" aspect of postscientific-resource-management thus correspond to the advantages I have touted in this book. Note, especially, that the advantages under item 2 in her list are improvements in the public's scientific understanding of environmental problems and "transformations" in the understanding of ecological systems and their complexity. She thus provides strong support—consistent with the findings of Sabatier and associates—that participatory efforts enhance social learning, increase scientific understanding among the populace, and contribute to the essential role of the more dynamic aspects of evaluation of environmental policies. If adopting more complex and holistic models of natural systems extends concern to longer time frames, perhaps this expanding awareness could lead to a transformation in public thinking in the direction of sustainability.

Layzer's critiques of EBM now appear in a new guise. The positive aspects of landscape-level adaptive collaborative management in creating social learning and improved public attitudes toward environmental protection have emerged from the twists and turns in management approaches, as environmentalists have struggled to design processes that could transcend the problems of SRM and lead to greater environmental progress beyond the simple passage of regulatory legislation. Layzer's critique of EBM can now be seen as looking more carefully at what works best within this broadly accepted idea of more collaborative and holistic management. She finds that EBM processes have been less successful than other processes, such as the GREM processes she uses as her control group. While one may be justified in doubting that Layzer's work provides the last word on this subject, one can see her careful data-gathering and presentation of her findings as offering finer-grained determinations of what works better among the many styles and approaches, which includes all seven of Layzer's cases plus the hundreds of watershed partnerships identified by Sabatier and associates.

Layzer raises two important cautionary notes and testable hypotheses:

1. Conventional political institutions—legislatures, local governments, and the courts—may be more effective in protecting the environment than are new institutions created by fiat and designed to achieve solutions through collaborative agreements.

2. Engaging all stakeholders in the early, problem formulation and goal-setting stage of an initiative may weaken resulting environmental protections, since politically powerful interest groups, especially development advocates, may lead to reformulation of the problem as one of maintaining the status quo while avoiding legal liabilities. If stakeholders believe decisions will be made by consensus, they perceive that they have veto power and can therefore prohibit consideration of solutions strong enough to protect the environment.

I can, then, accept the full range of data presented by Layzer as broadly supportive of the adaptive collaborative approach. At the same time, she is expressing concerns about the effectiveness of ad hoc, consensus-based institutions developed to address landscape-level problems, especially when they have broad scope and top-down origination. Here, we can—perhaps with a dash of skepticism, since some of her arguments clash with the broader findings of the Sabatier group—see Layzer as carrying forward the important task of empirically determining, among processes that are adaptive and collaborative, the exact factors that support the success of such processes and the factors that impede environmental progress.

Having so concluded, I must examine two important aspects of the way Layzer understands the characteristics of the seven cases she considers. First, consider her criticism of processes that adopt a "consensus" approach; she claims this can lead to powerful interests being able to veto strong plans. This apparently conflicts with findings of Sabatier and his group, yet the apparent conflicts may be based on a confusion of terms. Sabatier and his colleagues found strong, positive evidence that consensus-based management has had positive impacts on the adoption of policies for protection and restoration of damaged watersheds. These findings seem to clash with Layzer's findings, but in fact, they use the term "consensus" so differently that their apparent disagreements fade. As noted, Layzer claims that participants in consensus-based processes will be able to exercise veto power, employing a strong sense of consensus that requires 100 percent agreement among all stakeholders. Sabatier's group, by contrast, found positive evidence of effects of consensus-based decision making by understanding consensus in a much less restricted manner. Of forty-four watershed partnerships studied, they classify 93 percent as using a "consensus-based process." Elaborating, they described these processes as requiring a unanimous vote (11 percent), consensus of main participants (11 percent), or informed consent (participants can veto decisions they strongly oppose, by 82 percent). Other partnerships required majority or supermajority votes (14 percent; Leach et al., 2002, p. 7).[2] Layzer's indictment of consensus-building approaches is somewhat weakened if one recognizes that groups can set their own rules and that, in practice, decisions are often made without full-blown consensual agreement. Further, even if action is sometimes delayed indefinitely by vetoes exercised, this need not undermine the gains such processes make in shifting public understanding and perspectives.

Second, Layzer's critique of EBM also suffers from a confusing approach to the problem of scale. As noted above, Layzer selects her seven comparative case studies as all adopting a "landscape" scale and she says that the advantages of collaborative management result from addressing problems more holistically, including whole landscapes. But this makes her criticism of EBM somewhat confusing: since she seems to attribute landscape scale to all her seven cases, that variable apparently drops out of the analysis. Nevertheless, she ends by criticizing EBM approaches because they seem to bypass possible local avenues that were pursued with success in the three comparison cases (Pima County, Kissimmee restoration, and Mono Lake), which had a much smaller geographic extent than the four test cases (Austin's Canyonlands, San Diego Habitat Conservation Plan, Everglades, and CALFED).

To clarify this situation, it is useful to distinguish two ways in which one might describe the "scale" adopted in an adaptive, collaborative process. First, one can think of the scale of a problem as simply the spatial extent of the system to be studied and/or managed. So considered, the three comparison cases do, in fact, have considerably smaller ecological footprints than the four test cases, although it is not inaccurate to say that all seven, by comparison to the atomistic approach to problems by the scientific resource managers, adopt a "landscape scale." The EBM cases are in this sense more holistic and involve a broader landscape and probably more heterogeneity of lands and land uses. But since Layzer describes all seven cases as adopting a landscape scale, this sense of spatial extent cannot be used as a variable to explain the differential success among her seven cases.

Regarding her second important characteristic, one can think about scale as applied to a process not as spatial extent but rather as a perspective on an open system. Here, the relevant variable is the perspective from which a landscape-scale chunk of the environment becomes the focus of a "public" that, in turn, initiates a process to address problems from that perspective. In this book I have argued for the importance of adopting a local perspective, but one that recognizes that local systems are embedded hierarchically in an open system of expanding physical scales. One can suggest, then, that it may be this second sense of scale—the level of the socio-physical-political system at which individuals and groups decide to organize and begin the process of protecting natural systems in their own backyard—that is explanatory in Layzer's cases. This sense of scale may be useful in differentiating Layzer's positively described three comparison cases from her four test cases of which she is critical. In the Pima County, Kissimmee, and Mono Lake cases, the main impetus for organizing and seeking political actions originated in local communities that were inspired by local leadership.

On the other hand, each of the EBM cases was originated by the federal government, usually in an attempt to manage legal issues that arise over much larger landscapes and typically in situations in which local governments and institutions lack jurisdiction to manage the whole system embodying the problem to be addressed. A dramatic case of this effect will be discussed in section 9.4, where the habitat plan required to support several endangered species would require adjusting water removals from the upper and lower Platte River systems, which extend across three states and serve thousands of water users. The federal government addressed this complex, seemingly impossible problem by offering to facilitate a three-state solution to the problem of downstream habitat loss, because any solution clearly required cooperation across the entire Platte River watershed.

Layzer's results, then, may miss an important integrative insight that is implicit in her data. The large-scale problems addressed by EBMs can be seen as occurring at a larger, more inclusive scale than the problems addressed by Layzer's comparison cases. In these latter cases the effort was clearly "bottom-up," and the improved outcomes of the three comparison cases may well result from the advantages, touted throughout this book, of addressing local problems and then scaling up. Thinking hierarchically, one can envision large-scale management problems such as protecting the Everglades as occurring at a larger scale that encompasses larger and more complex systems than can be considered "local" in comparison to the Mono Lake or Sonoran Desert cases. This apparently definitive contrast between bottom-up and top-down approaches will be developed more fully in section 9.3.

The four EBM cases in which the federal government stepped in and initiated an overarching "ecosystem-based" process were situations in which either local governments had failed to begin any protective efforts or the problem loomed so large that locally developed policies proved inadequate to broader goals such as protecting biodiversity. Applying the hierarchical understanding of scale, one can then suggest that Layzer is comparing local processes with processes that cannot be—or at least have not been—addressed as local problems. EBMs should thus be seen as attempts to address problems that are, in fact, at least an order of magnitude larger than Layzer's three control cases. Layzer, in this interpretation, is comparing systems that are not actually comparable.

The control cases arose in situations where local people could effectively address the problems faced; Layzer's EBM test cases, by contrast, are too large scale for truly local processes to be successful. This reasoning suggests that adaptive processes are best initiated locally and that, in situations where systems of great geographical extent are in danger, it may not be surprising that externally imposed efforts are necessary but also prove less successful. If one treats the four EBM cases as addressing complexes affecting multiple local situations, it can be suggested that, in such cases, it may be more appropriate to initiate multiple local efforts, recognizing that these locally incentivized processes can be important drivers and, in some situations, multiple local processes can come together to create larger, cooperative groups that can carry their concerns to larger and larger scales.

Seeing Layzer's two categories of cases in this way suggests that, while locally driven processes are ideal, the larger-scale, more complex situations that occur at a larger scale may not, in fact, yield to impositions of process from central authorities. By thinking of local efforts as the primary scale for adaptive management processes, one can thus explain many of the differences that Layzer observed between the GREMS and the EBMs. She is certainly correct that the large-scale, top-down cases are more challenging; and in some cases it may be necessary to begin work at local levels and then scale up from these levels by forming federations of local processes to address larger, regional problems. Once we think of Layzer's two categories as addressing problems at different scales, we can anticipate that efforts initiated at a local scale may find allies in other local groups, may form a larger public, and thus may create interlocked, polycentric institutions in which the different levels support and complement each other. Since local origination and bottom-up management are ideal, it may be that larger-scale problems will need to be addressed by coalitions of local groups that form and then engage other local communities so as to deal with larger problems affecting them all.

It is easy to be cynical about cooperative processes and some of the claims made for them. Layzer offers careful analysis and debunks some claims of success, especially as just noticed, in the large-scale EBM cases. But if one looks more broadly at the empirical literature, if one views the whole range of watershed partnerships and other locally based organizations attempting to protect important places and resources, cynicism should fade. In fact, there are many examples of at least partially successful collaborative processes and also many examples where adaptive science is replacing uncertainty with scientific study and conflict with mutual learning. The empirical evidence here is strong: the landscape of resource use is dotted with examples of collaborative management projects of varying ages and levels of development. Further, these processes clearly outperform the now-discredited methods of SRM, especially in supporting social learning and in educating the public (Leach et al., 2002).

While one would like to think that all problems will eventually yield to the efforts of locally incentivized collaborative processes, it is obvious from even casual observation that, so far, the transition from old SRM thinking is incomplete and the path forward is not always clear. Yet this recognition only provides stronger motivation to continue the work of social scientists in their efforts to develop and evaluate democratic and cooperative pursuits of environmental protection.

The guiding light of this research lies in the work of Elinor Ostrom, her associates, and the other supporters of locally based, cooperative attempts to protect common resources. The good news is that empirical research is finding more and more cases in which cooperation has replaced conflict and in which locally developed and voluntarily followed conventions protect resources as well as entire ecological systems. Ostrom has also emphasized that such cooperation may involve not only local governance but also layers of polycentric institutions that, in the best of circumstances, can complement and support local control (Ostrom, 2007).

Still, Ostrom and advocates of cooperation are surely well aware that such cases are unusual and that, while one can cite many historical examples, it is devilishly difficult to create and nurture such a cooperative system in situations where most actors are already engaged in a competitive, selfish pursuit of individual wealth. These cautions, elaborated by Ostrom and her collaborators and students, pose an important challenge to the goals of this book. Does anyone know how to create a collaborative, cooperative process by which individuals and groups can come together and create locally effective processes capable of protecting natural systems and human access to them? Can one then knit these groups and processes into a multiscaled, polycentric system of processes capable of managing cooperatively at larger and larger scales?

To explore how these challenges are met in the real world, I consider two detailed case studies. Both examples have broad scope and have required multilayered understanding. The two cases also illustrate the importance and complexity of the scale issues—especially the question of top-down or bottom-up motivation—that will be developed in the next section.

9.3. Top-Down versus Bottom-Up Motivations

As one examines the empirical findings of social scientists who have studied the many recent collaborative efforts at managing resources adaptively, it is not surprising to note a mixed scorecard. Throughout the book, it has been recognized that in many situations entrenched powers favor the status quo because of advantages they currently have and that truly cooperative approaches may not be easily instituted. Nor is it easy to identify which cases are "ripe" for a cooperative process—for obstructionists do not usually announce their intention to dissemble and undermine a process. Nevertheless, there certainly are a number of functioning, effective ongoing processes that create trust among individuals who once acted as enemies. Based on this unquestionable fact, and with the intention of learning more about the details of how such processes work, I next present two case studies that have been carefully described and that allow evaluation of the processes involved: first, a compact formed among three states, two federal agencies, and countless water users in the Platte River basin to develop a habitat recovery plan for four species; and, second, a multidecade public process of discussion and scientific study that has led to new understanding and new management plans for the Chesapeake Bay Estuary on the eastern seaboard of the United States. The two cases differ in many respects, so I will discuss one key difference between them before advancing to introduce the two cases in sections 9.4 and 9.5, respectively.

As noted in the previous section, it is important to recognize a major difference in the motivations driving individuals toward cooperation: one "top-down," the other "bottom-up." In both cases, it can be expected that individuals will turn to cooperation when and if they are convinced that they can expect (or hope for) a better outcome as a result. This is the *incentive* to cooperate; and it turns out that the particular source of this incentive is important in shaping the kind of process and the kind of agreements expected.

In some cases, the incentive originates locally among affected individuals who conclude that continued competition will lead to worse outcomes than would cooperation. In these "bottom-up" cases, the motivation is largely from directly affected parties who recognize that the perceived problems result from their actions and those of their cohorts. These cases represent some of the more interesting, because self-motivating, examples of public reactions in the face of threats to common resources. While it may not be purely bottom-up in nature, improving management of the Chesapeake Bay region has enough bottom-up aspects so that it can serve as an example of bottom-up-motivated processes.

There is, however, a second type of case in which an overarching entity, such as the federal government in the United States or the European Union establishes a general policy that must be accommodated by local resource users. Some of the best examples of the top-down motivation and its downward impacts, as just noted, can be seen in the United States in the application of the US Endangered Species Act of 1973 (ESA). One example of such a case is the process by which the Clinton administration instigated the Northwest Forest Plan by wielding the ESA as a means to impose serious sanctions if cooperation failed. In this 1993–1994 case, the federal government forced resource users to the table by threatening to visit disastrous repercussions on them if they failed to reach a cooperative agreement.

The act was also the driving force behind one of the more complex and remarkable recent breakthroughs in adaptive management, the Platte River Habitat Recovery Program (concerning habitat for the whooping crane and three other species). This plan was formally instituted in 2007 following more than a decade of negotiations and conflict. One participant made the following assessment of the role of the ESA: "I think the ESA is a remarkable piece of legislation. . . . It's the one federal environmental statute that deals with scientific uncertainty and makes it clear that the species will not bear the burden of scientific uncertainty" (Daniel Luecke, quoted in Freeman, 2010, p. 30). This case will be the subject of section 9.4.

The outstanding aspect of these top-down motivated processes has been referred to as "the hammer" (Freeman, 2010). This refers to a centrally administered law or policy that can be used to force resource users into a process of deliberation and compromise by threatening to create a disastrous situation for them if they fail to reach a collaborative agreement.

The two types of cases and the incentive structures involved differ significantly. In top-down cases, the hammer comes down from the outside when local failure to cooperate is threatening values enforced at a state or federal level. By threatening to wield the hammer, central governments can force local competitors for resources to the negotiating table by warning of a very bad outcome unless they come up with a cooperative plan. In bottom-up cases, by contrast, incentives to change resource use policies emerge from a growing local consciousness by individuals and groups that their actions are creating threats to their own way of life.

In addition, the two cases stand at quite different stages of development, which explains why I will examine the Platte case first and the Chesapeake case afterward. The Platte River case, which involved strong controversy and conflicting forces from the beginning, required a long period, beginning in 1997, for constructing the rules by which the negotiations would be carried out before matters of substance could be addressed. The Chesapeake Bay case, on the other hand, while generating a great deal of controversy since water quality issues became public in the 1970s, moved forward because of a widely shared view of all involved that further degradation of the water quality in the bay was unacceptable. The two cases, then, not only show a contrast in the political situations involved but also illustrate two different stages of the process: The Platte case illustrates the painstaking efforts necessary to initiate a complex institutional response to a large-scale problem, while the Chesapeake process, having operated for over forty years, illustrates some of the long-term problems of implementation.

9.4. Endangered Species and the South Platte Water Plan

Like others, they did not know the value of a plover, a tern, a whooping crane or a pallid sturgeon. Whatever that value, it is not to be measured in market exchange of private goods. (Freeman, 2010)

Academics generally learn about adaptive collaborative processes only after they are well under way and their actions make news. As a result, most of our knowledge about how group processes are forged is based on the memory, perceptions, and viewpoints expressed by a senior member of a group expressed in interviews years after the fact.

Recently, a new publication supplements this after-the-fact, snapshot research with a detailed, event-by-event account of a complex collaborative process undertaken in the Platte River region in the center of the continental United States. Author David Freeman, an environmental sociologist and expert on western water politics, undertook to trace (by attending meetings, taking notes, and interviewing participants) the process by which the states of Colorado, Wyoming, and Nebraska addressed a conflict over how to accommodate the federal requirements of the ESA.[3]

Freeman's book provides a detailed account, assembled from experiences and notes taken over thirteen years, of the many conflicts and compromises required to achieve a basin-wide agreement. The resulting agreement involved three states, two federal agencies, several powerful water management districts, and countless water users. The agreements committed all involved to engage in a process of adaptive management—designed, through a combination of procedural agreements, substantive agreements, and agreements, to submit remaining disagreements to scientific study—in order to learn how to provide critical habitat for four species listed under the ESA. The account Freeman wrote, including anecdotes, data, and useful theory, provides a wealth of detail about how one particularly long, complex negotiation lurched toward an agreement, eventually accepted by all parties in 2006 and signed into law by President George W. Bush in 2008. A contentious group of water users, state officials, water managers, and politicians, against all odds, agreed to work together to improve the habitat of the whooping crane, the piping plover, the least tern, and the pallid sturgeon—species whose habitats in most cases existed hundreds of miles away from the homes of the stakeholders.

It is hard to think of a more complex—indeed, "wicked"—management problem than this one, and we are fortunate to have Freeman's detailed account. I will use that account to briefly set the context for this complex problem, summarize how the problem was resolved (at least for now), and draw some lessons about the role and importance of adaptive management both for resolving unresolved questions over time and for holding agreements together. The question, provoked by a US Fish and Wildlife Service (FWS) finding of jeopardy in 1994, was whether participants could stitch together a cooperative agreement on means to protect and enhance the habitat for the endangered species. The answer, as it turned out, required more than a decade of conflict and numerous partial solutions before an agreement for a first-phase management plan was put in place. The case of the South Platte River negotiations and agreement illustrates both how difficult the problems faced can be and also how persistence in pursuing a shared goal of management can address problems of extraordinary complexity.

The South Platte River originates in northeast Colorado, picks up further flow from southern Wyoming, enters Nebraska, and eventually joins the North Platte waters from Wyoming and northwest Nebraska near Lexington, Nebraska. These flows, originating in snow melt and mountain precipitation on the eastern slope of the Rocky Mountains, have supported extensive crop irrigation and economic development in all three states. Distribution and delivery of this water to diverse users from agricultural irrigators to growing cities—a task that cannot be accomplished by individuals—has led since European settlement to vast and powerful social organizations for the allocation of water. Until recently, these organizations and their behavior have been driven by stakeholder groups devoted to protecting the interests of their members, with strong support in their state legislatures.

Over the decades, growing demands, negotiations, and court decisions had, by the 1970s, created a complex tangle of water law, water rights, and organizations to administer and protect hard-won access to water, essential to life and development in the semiarid Midwest. In wet years, these organizations merely administered rules, since most users received adequate supplies. But occasional droughts provided reminders that, on average, there were multiple suitors for every drop of water. In many of the areas along the Platte, those water allotments that were granted outpaced the flow available, so managers were forced to allocate water according to priority of water rights—usually determined by seniority, or the date of first legal appropriation.

The system, as it had evolved by the 1970s, was driven by self-interest and prodevelopment policies in all three states. The existing regime had led to serious degradation of the natural flows of the South Platte River commons, to the extent that the river's users were living a classic example of the tragedy of the commons. When the self-interested users of water had fulfilled their demands, not enough water was left in the river to support wildlife habitat on the main stem of the Platte River downstream of its confluence with the North Platte River. There was no shortage of conflict and competition both within states (between groundwater and surface water users) and across states (as Nebraska consistently argued that the other states shorted them in their allocations), but these ongoing controversies within and across states were overtaken in the wake of passage of 1973's Endangered Species Act and subsequent events. Section 7 of the ESA requires that any federal agency, or any activity regulated by the federal government, may not further endanger a listed endangered species; this requirement was interpreted to apply to

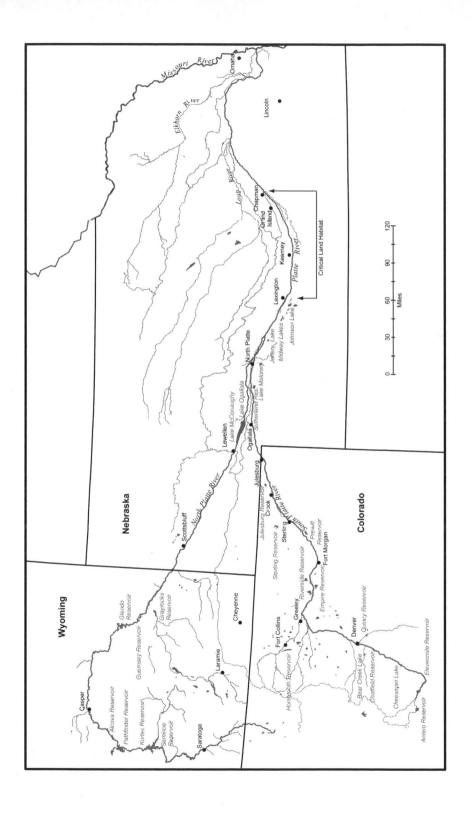

water policies and allocation. Section 4 of the ESA establishes an obligation to designate and protect "critical habitat" for such species, interpreted to require protection of sufficient land, water, and airspace to allow the species to survive and thrive. When the US Fish and Wildlife Service (FWS) judges that some action falling under the regulatory purview of the federal government further endangers an endangered species, it issues a "jeopardy" opinion.

In 1978, the FWS designated a fifty-sixty-mile-long and three-mile-wide stretch of the Platte River in Nebraska as critical habitat for the whooping crane and three other species. It argued that late-summer flows were inadequate for these species and that the leveling out of flows by dams, which dampened spring pulse flows, was changing the habitat by narrowing the river and increasing vegetation on the banks. (It had been determined that whooping cranes—to forage comfortably with enough sight distance to detect approaching predators—require a wide, open river braided with sandbars, a condition dependent on spring flood waters that rip out encroaching vegetation.) Based on this judgment, the FWS issued a jeopardy opinion with respect to upstream water use, including virtually all water projects in the Platte River basin. Under terms of the ESA, issuance of a jeopardy opinion requires that the out-of-compliance entity must create and submit to the agency for approval a "reasonable and prudent alternative" to its current activities that can be accepted as such by the FWS.

These events created a situation in which almost every water-use project in the river basin would require approval by the FWS, based on its estimate of a very large deficit in flow into the critical habitat downstream. Even though water users never agreed with the estimates of deficits set by the FWS, the agency had the authority to enforce the ESA. This situation created a terrible dilemma: without an agreement, every user of the river upstream from the critical habitat would have to face the federal agency individually and present a reasonable, prudent alternative to its own current actions, one that could convince the agency that the alternative action would no longer jeopardize the species. Water users either had to face the federal bureaucracy alone—a frightening and costly alternative—or had to band together and seek a basin-wide solution through negotiations with the other states and the federal government.

Figure 9.2. (*facing page*) Map of the Platte River system: This map shows the complex river system that had to be managed to protect endangered species downstream in the main stem. Note the area in which habitat land will be purchased or limited in uses by easement to create critical habitat. From "Platte River Recovery Implementation Program: Final Environmental Impact Statement," volume 1, US Department of Interior, April 2006, modified after Freeman, 2012. Courtesy of David Freeman and University Press of Colorado.

Perhaps the most salient aspect of this particular case, aside from its extraordinary complexity, was the dominance of the top-down aspect of the incentive structure. Water users, like most citizens of western states, already had a very suspicious attitude toward the ESA, seeing it as another example of the federal government's overreaching into state and local matters. In the Platte River case, federal intrusion was especially blatant: the FWS, on the basis of a biological opinion, wanted to regulate and limit water use in three states, including two huge river systems (the North and the South Platte basins) to create improved habitat for species federally listed as endangered, habitat that in some cases was hundreds of miles distant from the water users' locations.

It was by no means a one-sided conflict, as the states and water management districts had legal precedents, court cases, and strong local institutions that favored the status quo, plus they had the ability to delay and drag out the process for years. In the end, however, the legal situation was clear: the FWS held the trump card. FWS had clear jurisdiction: on rendering a jeopardy opinion, and backing it up with a science base considered adequate by reviewers and the courts, it could demand that all users comply by proposing and implementing a reasonable, prudent alternative to their existing policies, procedures, and actions. If every user had to face the FWS individually, the costs and the disruption of normal activities would be enormous. This specter caused water users to seek some kind of regulatory certainty—some agreement within which they could plan and work, without fearing constant challenges from the FWS.

It was this massive hammer, wielded by the FWS, that brought participants to the table and kept them there despite apparently unresolvable problems. It was also true that the FWS could not be certain it had—or could find—resources sufficient to deal with so many users in complex individual negotiations. So the battle began, in 1978, when the FWS designated critical habitat on the Platte River (below the confluence of the North and South Platte Rivers). Over the next few years, five more species were declared endangered in the central Platte system. In short order, FWS issued jeopardy opinions for virtually every proposed expansion in the use of river water. It was clear that the barrage of jeopardy opinions would both freeze development activity in many situations and close down many businesses based on water withdrawals. Subsequently, as jeopardy opinions proliferated, the problem was subject to two intense studies: an environmental impact statement (EIS) was undertaken under the auspices of NEPA (the National Environmental Policy Act), and FWS undertook a biological study of threats in search of reasonable and prudent alternatives. Meanwhile, the tension built and pressure to respond to the jeopardy opinions grew. The feared outcome—that each and every water user would have to respond to the FWS—was delayed in hopes that a basin-wide solution could be found. At last, in 1994, the three involved states signed a memorandum of agreement with the FWS. The states pledged to make a good-faith effort to find a comprehensive solution to the complex upstream problems that threatened species in the central Platte basin.

The FWS moved slowly, hoping to achieve a cooperative agreement, because it too had reason to be terrified by the possibility that each case would be addressed—and probably litigated—on its own. Under the circumstances, the FWS chose not to press the issue immediately, so it granted temporary continuance of activities, provided that serious and fruitful negotiations were undertaken by all Platte basin interests, with the intention of proposing and seeking approval of a basin-wide solution that would mitigate the jeopardy findings (Freeman, 2010, pp. 33–34).

This crisis led in 1997 to a Memorandum of Agreement, signed by the governors of the three states and by the Secretary of the Department of Interior, pledging a serious, good-faith effort to put together a cooperative process by which the downstream critical habitat of the listed species would be protected and enhanced, jeopardy opinions would be reversed, and water users would be able to gain compliance with the ESA as a group rather than as individuals.

The story Freeman tells, of state-by-state negotiations (bathed in ancient mistrust over water sharing across state lines), of emergent problems within states (including stirring up controversies between surface and groundwater users), and of deep suspicions of anyone associated with the federal government, is far too complex to recount in any detail here. We have to be content with mentioning a few salient characteristics of the negotiations and with drawing some lessons and clarifications from the process and its successful outcome.

Beginning with the 1997 Memorandum of Agreement, the processes of negotiation unfolded: the Department of Interior issued a fifteen-page memorandum outlining program goals for all participants; the department formed research relationships with the states; and a Cooperative Agreement Governance Committee was formed. As Freeman tells the story, negotiations proceeded in fits and starts, with each stage of the negotiations facing what came to be called "skunks"—issues that seemed to represent impossible clashes in perspective and attitudes. As each skunk was "deodorized" in one way or another, yet more skunks emerged. Understanding how progress was obtained is important because it illuminates the role of science and faith in adaptive management in converting skunks into empirical hypotheses that could be tested. As it turns out, adaptive management and the tendency to turn disagreements into testable hypotheses saved the process at several key points. The basic idea can be grasped by looking at how two, apparently intractable, problems were managed by setting up scientific studies to resolve basic disagreements among participants.

In addition to the central controversy over how much water would remain in the river when it reached critical habitat, there were also serious disagreements over what land to protect as habitat and how to protect it. The two challenges—first, of agreeing on a policy for land acquisitions and, second, of finessing the FWS's controversial pulse flow plan (deemed essential to maintain vegetation-free sandbars)—will be briefly described as illustrative of the difficulty, but also the possibilities, of managing wildlife habitat cooperatively even under hostile conditions. These two aspects of the planning process illustrate the importance of top-down pressure in maintaining the negotiations over multiple years and in many rancorous fights, and also how adaptive management can sometimes fill gaps between negotiators who are stuck some distance apart in their positions.

9.4.1. First Challenge

The FWS plan for developing critical habitat included regulating water flows, but it also required dedicating a large amount of land along the river as protected areas and, while it realized it would not be able to buy only ideal habitat in large chunks, FWS was committed to limiting purchases and other arrangements (such as purchasing development rights on private land) to large, contiguous areas that were also connected by corridors. Local governments, however, were loath to give up the potential tax revenues from development of prime riverbank parcels, so they lobbied to include in the acreage counts of protected lands purchases of sandpits—artificial ponds created by mining gravel—where the endangered species had been spotted. The FWS initially refused to use any program money to purchase such smaller, usually disconnected, parcels, so state and local participants held out for using this low-value land as a substitute for large, developable parcels along the river. The FWS then argued that just because a species is found in a particular place does not imply that that place is ideal or near-ideal habitat.

Shortly after the formation of a steering committee, the issue of choosing a policy for land acquisitions was separated out from the overall controversy and a land committee was formed to explore and develop policy with regard to how land designated as critical habitat would be protected. It is widely acknowledged that policies and decisions regarding land purchases are best done in small, nontransparent groups acting confidentially, at least to some degree; if preferences for purchases become known, speculators will drive up the value of the best parcels. So the FWS was in the difficult position of having to defend a plan for the purchase of land that was too vague to indicate what lands or even types of lands would be purchased. Further, it was walking into a hornet's nest of antifederal attitudes that was stirred up when the organization Nebraska First, already upset about attempts to more tightly regulate groundwater use, took an uncompromising position against both the land acquisitions and any attempts by the federal government to control land uses, which is a matter usually reserved for local decisions. Through 1997 and 1998, the meetings of the land committee were unproductive, as protesters, arriving in busloads organized by Nebraska First, shouted down any moderate voices and even prohibited the committee from setting rules for proceeding. In fall 1998, meetings were paralyzed by hostility as Nebraska First members, certain that the federal government was involved in a plot to grab their land, shouted down all speakers and intimidated participants; many observers worried that violence would break out as farmers, who fed on rumor and assumptions rather than accurate information, refused to even read documents or listen to explanations of the program.

The situation seemed hopelessly polarized, but eventually things settled down somewhat; Nebraska First agreed to discontinue busing members to meetings and to send just five representatives, who could speak, in turn, with other participants. By the end of 1998, the land committee had organized itself, chosen members, and was able to focus on its challenges. This progress, however, only focused attention on the question of how to describe adequate habitat. The FWS started from an abstract ideal representing a "perfect habitat" for the species in question and, while it was willing to accept that compromises would have to be made, it held out, at a minimum, that all protected lands should be contiguous or connected to other habitat, thus creating a few larger reserves. The range of policies it was ready to consider most assuredly did not include sandpits and other noncontiguous wet areas. Defining critical habitat thus came down to an impasse: the FWS would not yield on the question of discontinuous parcels, while locals refused to define adequate habitat as counting only prime, contiguous parcels.

It had been decided by the governance committee that the overall plan for creating critical habitat would be broken into increments, the first of which was to be a plan for the initial thirteen years, expected only to make progress toward the distant goal of having adequate habitat and delisting of species, not to achieve full success. Within this framework, the FWS agreed to set a goal of purchasing, in the first increment, 10,000 acres of adequate habitat for the four endangered species, and it held out for the best possible quality for all these acres. Local forces elevated this issue to a breakpoint, and it appeared that negotiations would dissolve. The two sides dug in over the question of whether the sandpits could count toward the 10,000-acre goal. The FWS refused to agree to any plan that mentioned sandpits explicitly as possible habitat; locals representing users, in turn, would not agree to a plan that limited purchase and protection options to some bureaucratic ideal of "perfect" habitat.

Attacked relentlessly by the states, the FWS made a conciliatory move; it added a clause to the agreement that allowed the purchase of up to 800 acres of "noncomplex habitat lands" that would count toward the 10,000 acres. What made this compromise acceptable to both sides were, first, the explicit admission by the FWS that the states' contention that sandpits might be adequate habitat could be true and, second, an agreement to engage in adaptive management by studying the effectiveness of the noncomplex habitats to determine whether they should be counted as adequate habitat. So, in effect, the two sides of this controversy were able, for the time being, to disagree about the efficacy of sandpits as habitat and to protect some sandpit sites in order to study them during the first, thirteen-year increment. In this way, an apparently irreconcilable difference affecting policy was finessed by substituting experiments for immediate agreements. Adaptive management proved its usefulness in this case by transforming apparently ideological differences into a testable hypothesis about the exact habitat requirements of important endangered species.

9.4.2. Second Challenge

Much later in the process, leading up to an agreement on managing water to mimic spring runoffs eventually signed in 2003, adaptive management again bridged a gap in the understanding of river behavior. Initially, the FWS mainly advocated increased water flows into the proposed critical habitat, but as time passed and more studies showed how whooping cranes and other species used the river, it began not only to advocate for a critical volume of water to be delivered but also to insist that some of that water be timed to create a pulse similar to spring snowmelt. Their studies seemed to show that holding back water in reservoirs and meting it out precisely as it was needed for agriculture and other uses (mainly in mid- and late-summer) had reduced spring flows, which had the effect of narrowing the river as vegetation encroached from either side and onto the sandbars.

State representatives never accepted the FWS theory supporting management of pulse flows, but FWS representatives came to believe that the program could only succeed in improving downstream habitat if it could manage spring pulse flows. Again, a seemingly unbridgeable gap existed, both in the policies advocated by the FWS and state and local governments and also in the science used by the federal government and other participants.

Once more, resorting to adaptive management saved the process from collapse. Through complex negotiations it was agreed that an "environmental account" would be set up in one of the large reservoirs on the Platte. This water would be saved during wet years and held back as a reserve that could be tapped by FWS agents to create a surge mimicking spring flooding for one to three days each spring. Once this was accomplished, the empirical question of whether such a surge would actually widen the river and increase habitat, as predicted by the FWS, came into focus. Unfortunately, this hypothesis was untestable at the time decisions had to be made because there existed a "choke point"—a narrow stretch of river that intervened between the environmental account water stored upstream and the location of the critical habitat downstream—and water released to form a surge would back up at the choke point, spoiling the experiment. Lacking a way to support its hypothesis, the FWS was unable to convince the states and local users to save the water for surges, but it was unwilling to accept a plan with no spring surges for fear that their lack would undermine all the other advantages gained by the program.

With negotiations on the rocks, the participants again turned to science. Realizing there was no feasible way to test the hypothesis that pulses improve habitat at the time, and that finding solutions to moving the water downstream in the proper amounts to create a pulse would require time, it was agreed to engage a scientific research team to study the problem. Their report described, using computer models, several possible ways to construct and execute a pulse flow program. The parties, exhausted by conflict, finally agreed to sign the agreement, pledging to study, on the basis of the research team's identification of thirteen possible solutions that could be tried, how to create the conditions necessary to test the pulse flow hypothesis in the first year of the first increment. Once again, by recasting a matter of disagreement in policy as a hypothesis about the effects of experimental action, a hard-won agreement was eventually forged.

The Platte River case is highly unusual, first because of its extent over much of three states, and second with benefits of management expected to occur far downstream of the individuals and groups that would be asked to sacrifice some of their status quo rights. It is also distinctive in the extent to which the process was driven from the top down. Water users and their state protectors began with no incentive to agree to limits on their current behaviors; it seemed that they cared little for endangered species, doubted their activities really affected whooping cranes, and began with deep suspicion of the federal government. It was only the insistence of the FWS on its mandate to identify and protect endangered species, and its dogged determination not to allow outcomes that could not be supported by science, that kept the participants engaged. The hammer—the threat that most users of Platte water would have to address jeopardy alone—kept participants at the table. Ultimately, however, it was deference to science—leaving open controversial questions that could be empirically tested as the process moved into its first increment—that pushed the process toward agreement.

There is no denying the difficulty of addressing complex problems like the Platte River case; it took over a decade, countless meetings, many metaphorical logjams, and deodorizing one "skunk" after another, but an agreement to continue to negotiate and to persist in studying the systems under management is in place, and now it is up to everyone to learn from this and other important experiments in holistic, adaptive management.

Directly preceding this section, I included David Freeman's observation that "Like others, they did not know the value of a plover, a tern, a whooping crane, or a pallid sturgeon." Just before making this remark, he quoted an observer:

> Important components of a viable society cannot be captured in marketplace exchange and by their summation in gross national product accounts. At some point, in recognition of this truth, people must mobilize themselves through their representative governments and local common property management organizations to address matters by investing in the commons. In doing so we add to our collective capacity to govern ourselves in our landscapes.
>
> . . .
>
> To produce this new governance system for the environmental restoration and sustenance of this collective property, the players have had to agree to transcend and adapt their particular private and historical agendas. (Freeman, 2010, p. 10)

One might wonder, given the important contrast between top-down and bottom-up incentives driving programs, how one can determine whether a given environmental problem is more likely to be improved by appealing to top-down, centralized power or whether solutions will bubble up from below once the public and stakeholders decide that the problem must be addressed. Recalling Layzer's distinction between EBM and GREM, and the suggestion that the difference may really be a manifestation of a difference in scale, one can see the glimmer of a solution. Today, apparently, GREMs seems to work well at small scales, where meetings and relationships are face-to-face and where maximal opportunities exist for individuals and groups to learn from each other. Let us informally call this the "place scale." A place, as noted earlier, is where the people feel a natural affinity to each other and to their land; a place is a community both of people—of all ages—and of wild species and their habitats. At this scale, individuals have a chance to discuss actions that can be taken as well as motivations and reasons for these actions, in a context that seems highly effective in creating social learning and social capital. Here, one hopes that the dynamic aspect of evaluation will come to the forefront, that local communities will learn how to make better decisions affecting their environment, and that making better decisions will guide such communities toward a sustainable path forward. At small scales, motivation can emanate from local communities, as is illustrated by Layzer's examples of protection of the Sonoran Desert in Pima County and of Mono Lake restoration. Top-down force may not be needed in such cases and may even be counterproductive.

By contrast, cases as large-scaled as Layzer's four test cases and as unwieldy as the Platte negotiations can be thought of as scaled up to a larger level. These cases are so large they will include multiple small-scale places, and here the problem is to find motivation to stitch together multiple, smaller places so as to address a larger problem too great in extent to be handled locally. Layzer's cautions about attempts at EBM inform us that, if adaptive collaborative management is going to scale up to address larger, second-generation problems, there will need to be some source of motivation to bring disparate interests together at a scale capable of addressing problems as big as the Everglades or the Platte River basin. At these scales, social learning is apparently attenuated and agreements usually are made because participants are convinced that the cooperative solutions available are the best they will achieve, not because of transformations in perspective or values. For the foreseeable future it would thus appear that up-scaling adaptive management practice will usually depend on top-down motivation by which centralized governments use a hammer to drive reluctant warriors into cooperative processes.

This speculative note highlights what may be the key insight that can be learned from our study of adaptive management of complex, many-scaled systems. Social learning through adaptive thinking and experimental management can move inhabitants of a place to a fresh awareness of their environment and the long-term consequences of their behavior. Scaling up these local successes proves more difficult as environmental problems progress from first- and second- to third-generation problems. However good citizens become at protecting their local places, there remains a huge challenge when communities of relatively enlightened inhabitants of a place learn, through love and science, to feel concern for the space around their place. Buffeted by second- and third-generation problems, even greater challenges arise of applying adaptive collaborative management at larger and larger scales. No simple recipe for success at these larger scales exists. But examples like the Platte River case suggest that, given a strong legal mandate and persistent pursuit of cooperative agreements, a process can, even against heavy odds, succeed in achieving cooperative solutions.

9.5. Thinking like a Watershed: Remapping the Chesapeake

Can large communities experience transformations in their consciousness— things that, more technically, can be called their "cultural model"? I answer this question, "Yes," and furthermore I can provide a detailed historical account of how the people of the Chesapeake Bay region on the East Coast of the United States, residents of five states and the District of Columbia, replaced one mental map with another, more holistic and more ecologically appropriate one. My answer (and this case study) relies on results of a National Science Foundation study of how systems deemed problematic are bounded in public discourse.[4] While the transformation involved multiple scales, including an important US Environmental Protection Agency (EPA) study of bay functioning, this case does in many ways exemplify a bottom-up process.

This case study illustrates most of the positive features of a successful public process that led to a change in both individual mental models and the collective cultural models. As my research team and I puzzled over problems of scale in the formulation of environmental problems, we decided to undertake a retroactive case study of the Chesapeake. Since our grant was only for three years, and processes of cultural change unfold over decades, what we could learn from a current case would be limited. So, we decided to study mental and cultural models of individuals and groups residing in the Chesapeake Bay region for a period starting at about 1970. The much-reported and much-discussed case provides an excellent example of how inhabitants of a large and complex region can, through a process of social learning, undergo a cultural transformation that literally changed "the place" where they lived. The achievement was dramatically and metaphorically expressed by Tom Horton, the long-time environmental reporter for the *Baltimore Sun*: "We are throwing out our old maps of the bay. They are outdated not because of shoaling, or erosion or political boundary shifts, but because the public

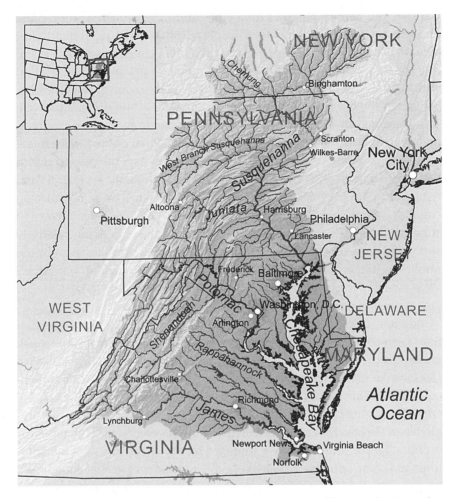

Figure 9.3. Map of Chesapeake Bay: Early concerns about pollution of Chesapeake Bay emphasized its direct polluting sources; once it was learned that a major threat was point-source pollution (such as nitrogen and phosphorous runoff from farmers' fields and suburban lawns), participants were forced to care for the entire watershed. Map by Knusser, Wikipedia (http://creativecommons.org/licenses/by-sa/3.0/deed.en).

needs a radically new perception of North America's greatest estuary" (Horton, 1987, pp. 7–8).

Between 1965 and 1995, Chesapeake Bay was transformed, not by natural forces or by human engineering, but through a collective act of social learning, which saw a collective shift of cultural and spatial consciousness that transformed the pollution problems in the region from estuary-sized problems into watershed-sized problems. Before this shift, concerns regarding pollution of the bay had focused on point-source pollution that could be attributed to sewage outflows from cities and large industrial sites, often allowing the identification of responsible culprits. In the 1970s and 1980s, spurred by both an important EPA study and independent research, a scientific consensus emerged that, while local, point-source pollution (i.e., that which originates from an identifiable and locatable source) remained important to the bay and its tributaries, a more diffuse and difficult problem was presented by the overnutrification of the bay from nonpoint sources, including in farmers' fields and suburban lawns. What ensued was a transformation of scientific and cultural models of the bay: the perspective of bay-area residents shifted to a larger, or rather additional, geographic and temporal scale. They began to see the whole watershed as an organism, a complex system that encompassed the entire watershed, spread over five states and the District of Columbia. A careful examination of this relatively successful process of social learning regarding the Chesapeake, whereby the public, politicians, and policy makers adopted a mostly shared, watershed-sized model of the bay, can provide clues about how to solve today's wicked environmental problems and achieve robust adaptation.

I do not, of course, claim that the problems in the Chesapeake region have been solved and that there are no longer disagreements regarding bay management. Indeed, several plausible studies have been highly critical of Chesapeake management and implementation of plans (e.g., General Accountability Office, 2005), but our focus was more narrowly on the process whereby scientists and the public came to have new mental and cultural models of the bay. They expanded their boundaries around the system, which set the stage for, and also was encouraged by, the emergence of new institutions and organizations that worked on different sides of several boundaries. So, we analyzed Chesapeake Bay management as a retrospective case study to better understand the role of scale and boundaries in the understanding of environmental problems. Our goal in the development of this study was to better learn how existing boundaries emerge, often without explicit thought, and how they solidify in the participants' consciousness. Then, taking on a life of their own, these boundaries shape the ways the participants conceive problems and limit the solutions open to consideration.

By retelling the familiar story of the transformation of Chesapeake Bay environmental policy from early attention to pollution and from point sources into a more comprehensive—and geographically much more extensive—understanding of non-point-source pollution, we offer this case as one in which a combination of science and a values-based awakening created an example of "social learning" at the level of cultural models. This retrospective case study shows that it is possible to transform the "mental map" of a whole region, creating a new cultural model of the bay system. In the process, deliberation and public discussion not only transformed individual and group consciousness, it also made it possible to begin to manage the system at a scale more appropriate to the real problem by focusing attention on the dynamics driving change. Eventually, this process led to the formation of new and more effective institutions that spanned a larger system, a process we came to call "macroscoping," which occurred in step with changing scientific and cultural models of pollution. This change in models, in turn, transformed the "problem" itself and also brought new players into the dialogue and deliberation.

From the 1930s and through the 1960s, concerns for the health of Chesapeake Bay ecosystems focused on several key point sources of pollution, including sewage outflows from cities and large industrial sites. This focus implicitly assumed a closed-system model of the bay system that included, basically, the main stem of the bay and a few sites on its tributaries. Beliefs about the "drivers" and the nature of pollution encouraged scientists and activists alike to adopt a geographic model of limited extent and to place an implicit boundary on the system that required study and intervention.

This "model" of Chesapeake pollution, which embodies a snapshot of accepted thinking about the problematic situation in the middle of the twentieth century, shaped all policy discussions and seemed to pit environmental activists against a relatively small group of polluters. This and other mental models, with their spatial extent limited to the impact area for particular, identifiable pollution sources, were relatively stable up until the 1970s. As long as public and expert opinion focused on these drivers and these culprits, mentally constructed geographic boundaries remained stable. However stable these mental and cultural models of the bay pollution problems were in the 1950s and 1960s, they were drastically overhauled in subsequent years and decades.

By the 1970s there were multiple danger signals that the bay was becoming polluted, even if it was unclear what that was driving the widespread changes in bay functioning, especially the increasing turbidity and consequent dieback of the vast underwater grass flats that formed the base of the bay's food-web. This case illustrates the important role of local leaders and their contributions to getting environmental problems onto the policy agenda. In this case, which mostly fits the bottom-up model discussed above, individuals and local groups began to agitate, complaining about increased pollution and falling populations of fish and shellfish. Criticism was heaped on known point-source polluters. Maryland State Senator Bernie Fowler did stump speeches all over the area in which he complained, "I can't see my toes!" A native of the area, he remembered when he and his family could stand in waist-deep water and dip crabs. But by the 1960s, Fowler could no longer see his toes in shallow water, and his complaint became a sort of rallying point: "Why can't Bernie see his toes?" And "What should we do about it?" This locally driven, bottom-up impetus forced the issues onto the agenda and, as is not unusual in bottom-up driven processes, the local outcry reached the EPA.

In the early 1970s, spurred by public outcry and by alarming scientific studies, the EPA undertook a detailed scientific study. It was learned that, while environmental monitors were paying attention to small-scale, local variables, a large-scale variable associated with a larger-scale dynamic—one driven by the total input of nutrients into the bay from its tributaries throughout the watershed—posed a slower-moving, but more profound, threat to bay health. Agricultural and residential runoff of nitrogen and phosphorous was causing algal blooms and anoxia in deep waters, causing increased turbidity and reducing submerged aquatic vegetation. The rich farmlands of Pennsylvania, the Piedmont, and the coastal plain all drain into the Chesapeake. To save the Chesapeake, it would be necessary to gain the cooperation of countless upstream users of the waters that eventually enter the bay—a monumental task, since Pennsylvania and the District of Columbia, situated upstream on tributaries, have no coastline on the bay and no direct stake in its protection.

This "transformation" of the bay from an estuary into a watershed was a cultural shift that occurred in a context of mission-oriented science; it was as much a process of transformation of public consciousness as it was a change in scientific understanding. It signaled a dramatic change in perspective that was driven by values—an outpouring of love and commitments not to let the bay further degrade. To address the problem of bay water quality, it was necessary to create a new "model" of what was going wrong with the system. The shift in models led to a public campaign, driven by the deep and varied values residents felt toward the bay, which was marked, for example, by the outstanding success of the Chesapeake Bay Foundation (a private foundation that advocates for, educates, and supports science to guide bay management). So, here is an example of a value-driven remapping of a complex natural system, how it works, and how pollution is being delivered into it. We can go so far as to say that a new "cultural model" was formed (Kempton et al., 1999; Kempton and Falk, 2000; Paolisso, 2002). Chesapeake Bay management, while not perfect, of course, has been a model of cross-state cooperation as serious steps were taken throughout the watershed to reduce non-point-source as well as point-source pollution.

How should we interpret this transformation? A scientific finding that the bay was threatened by processes outside its currently conceived boundaries, interacting with the strength of the love for the bay as a "place," created a new model that more accurately represented the problems of the bay and also expanded the sense of responsibility of its residents and users. The public understanding embraced the larger system, as their attention shifted to non-point-source pollution problems throughout the watershed. One could argue that it was values—the love felt for the bay—that was driving the acceptance of these models; it would be just as correct, however, to say that it was the scientific studies that analyzed the problem as watershed-sized that embodied this love in a new ecophysical model of the bay and its problems.

When Horton describes the change in cartographic terms, the underlying truth is that the shift to a watershed-sized model was the expression of an implicit value—a sense of caring for the health of the bay as a part of hundreds of millions of people's way of life. Values can be implicit in the choices of "models," scientific and mental, that people use to understand problems. The concern of individuals throughout the region came to express a community consensus in goals and values, transforming a local consciousness into a regional consciousness and broader sense of responsibility. Through social learning, the residents of the area discovered how to "think like a watershed" and began living in a larger "place" than before.

The retrospective Chesapeake case study reveals a process that remodeled the bay as a more expansive system and that also exemplifies many of the characteristics of a healthy, if not perfect, public process as advocated in this book.

- Local governments and local residents, who perceived negative changes in the bay and its functioning, spoke up, and local leaders, including a state senator, agitated for more studies and tighter regulation of polluters.
- Both experts and the public were involved; in this case, EPA scientists and consulting scientists acted in concert to gather available data on water quality and to commission studies where important data was lacking. The science was embedded and mission-oriented.
- The well-publicized scientific study by the EPA catalyzed a rethinking of the whole system, its boundaries, and assumptions about the scale at which impacts of local activities were playing out. There was explicit discussion of the appropriate model for understanding the "big picture" of polluting dynamics in the region.
- As scientific information was coming in, there were major institutional shifts and new contributions from organizations with a more holistic,

watershed-sized focus. These included, importantly, the Chesapeake Bay Foundation, located in the heart of the system on Smith Island. The foundation became a strong advocate of holistic, watershed-scale models and action. (One might argue that the foundation itself has served as a boundary object and a boundary organization.) Equally important, as it became clear that threats to the bay extended past any existing jurisdictions in the region, the five involved states and the District of Columbia formed the Chesapeake Bay Compact, a mostly advisory body that created an organization that overlaid state jurisdictions without threatening their sovereignty. (See Blatter and Ingram, 2001, for similar examples.) It took a combination of new institutions, as much as new scientific understandings, to reconstitute mental and cultural models as being inclusive of larger scales. But the compact demonstrates the possibilities and the appropriateness of ad hoc institutions that stitch together constituencies across political boundaries.

- The shift from addressing bay-scaled problems to addressing watershed-scaled problems went hand in hand with a gradual change in public discourse about management and the future of the bay. One of my graduate students, Paul Hirsch, did a content analysis of local newspaper coverage of environmental issues and pollution threats over a thirty-year period, which showed a close correlation between changes in public and media concern about particular pollution threats and the expanding scale of the mental and cultural models that became dominant in public discourse.

I am sure that the veterans of many years in the ongoing bay management controversies—those about fair sharing of burdens of management, about how to regulate non-point sources, and so forth—will find this list of successes far too positive in tone. Viewed from a day-to-day perspective, the process was actually messy, contentious, sometimes personal and vitriolic, and certainly not ideal. But if one looks at the big picture, over a period of around three decades, with support from the federal government, a whole region participated with policy makers, politicians, and the public in a remarkable change that can only be understood as a cultural shift in the models that give meaning to the events unfolding as a result of unintended consequences and management interventions.

If one looks at that big picture, rather than at the day-to-day squabbles and roadblocks, one sees that an initially inchoate concern for the bay was transformed into a campaign to study and to act to protect its processes. The campaign encouraged a process of gradual learning in which the bay, once the productive factory for a growing economy, became a watershed. We should not be surprised, then, that a metaphor lay at the heart of the

EPA literature explaining the change it thought it was seeing and supporting: the EPA published a booklet intended to explain its findings and their implications for managing the bay, called "Chesapeake Bay: Introduction to an Ecosystem" (Environmental Protection Agency, 2011). The preface to this document, signed by the director and the deputy director of EPA's Chesapeake Bay Program, says, "The Bay is, in many ways, like an incredibly complex living organism. Each of its parts is related to its other parts in a web of dependencies and support systems. For us to manage the Bay's resources well, we must first understand how it functions" (Environmental Protection Agency, 2012).

The transformation that took place across the Chesapeake region, then, is similar to Aldo Leopold's transformation: both involved a redrawing of the boundaries of a managed ecosystem, a redrawing explained and presented as a change, not in some specific belief but rather in the adoption of a new basic metaphor for understanding an entire system. Apparently, we need such a transformative metaphor to move up yet one more level to the global scale if we are going to comprehend and address third-generation problems, to which we turn in the next chapter.

CHAPTER 10

Addressing Third-Generation Problems

10.1. Scaling Up: The Emergence of Third-Generation Environmental Problems

Natural systems change naturally. They always have, and humans—just like other species—have always evolved in response to changes in their environment. The ecologist-forester Aldo Leopold was a master at adapting to different natural regimes in different locations, being acutely aware of the ubiquity of change and the need to constantly adapt. It has also been true that humans have always altered their environment, so changing in response to changing natural systems is partly a matter of adapting to our own adaptations. This interaction is sufficient to imply that, as we humans observe and evaluate change over time, we must take a dynamic, adaptive, and multiscalar view of how things change and how to evaluate such change.

So, understanding the "natural" relationship of humans to land requires a dynamic view, yet Leopold also learned in his experiment with wolf removals that some human activities not only change the local context in which those activities are undertaken but can also set in motion changes that radiate outward, affecting larger and normally slow-changing systems at much larger scales. Leopold learned that the longstanding truism to "expect change" only partly captures the quandaries of his day: he recognized that the kind of changes he observed on the wolf-free landscape were different from the limited effects of small groups of hunter-gatherers. He realized that the problems that were emerging in his day, problems affecting entire landscapes, reflected a new kind of challenge: larger systems usually assumed to be relatively stable have become vulnerable to human actions. Leopold explained these actions as follows: "Evolutionary changes, however, are usually slow and local. Man's invention of tools has enabled him to make changes of unprecedented violence, rapidity, and scope" (Leopold, 1939, p. 727).

So Leopold would probably not be surprised to learn that, today, changes at yet larger scales—almost certainly the result of human impacts on natural systems—affect systems with an even larger scale than the landscapes of the American Southwest. This problem of scale spillovers emerged for Leopold conceptually, as a result of his "successful" eradication of wolves. What he found was that, while killing a few wolves would have little effect on larger systems, extirpating the top predators set in motion cascading changes that began to accelerate change in normally slow-changing, environmental systems (Eisenberg, 2011). Leopold's successful program to remove wolves was so extreme that it did not result in a normal evolutionary response. These extreme changes were not "damped out" at the scale of populations, and could not be, because one of its key variables affecting populations had been loosed from its natural constraints. As the deer population exploded without a response from predator populations, impacts of his actions began to show up in accelerated change at the scale of the "mountain"—that is, the ecosystem. From this effect, Leopold learned that the experimental spirit in management must also be accompanied by a measure of humility as he came to regret the violence and irreversibility of his wolf-removal experiment.

I have carried forward Leopold's anecdotes about cranes and crane marshes and about wolves, hunters, and deer throughout the book because his essays express a seminal truth regarding human plans and natural systems. This truth has profound descriptive and scientific implications. The essays also evoke deep, almost visceral values that reside, according to Leopold, in a "higher gamut, as yet beyond the reach of words" (Leopold, 1949, p. 96). This combination, when mixed with the strokes of a master artist like Leopold, creates through metaphors a means to transcend increasingly inappropriate metaphors that exert a dead hand on current deliberation and thus to see problems in a new light. And this is what is required if one is to transcend the currently polarized dialogue about emergent—and urgent—third-generation problems like climate change.

10.2. Learning to Think like a Planet

Following the practice of learning from Leopold, it may be helpful to examine in some detail how he described his posttransformational understanding of ecological systems to a scientific audience. Perhaps clues can be derived by speculating about how he would go about adapting to the emergence of third-generation, global problems by showing how he adopted, deployed, and shifted metaphors when he was forced to rethink his original, utilitarian position with respect to wolves and deer. Today, puzzled and conflicted by climate change, humans need to undergo a similar feat of "scale-jumping" from our small, "chunked" view of the world that informs economic reasoning, to a more inclusive and integrative picture of how ecology and human activities play out on larger and larger scales. Like Leopold, the entire culture needs to supplement the short-term, small-scale viewpoint with a larger, more holistic perspective. Accordingly, it may be fruitful to examine in detail how Leopold explains his transformation to his colleagues in ecology and in the US Forest Service. We can learn, I think, from the way he weaves ecological threads, normative threads (economic and ethical), and metaphorical threads into his account of his transformation.

Leopold discussed the pattern of overpopulation following predator removal as one of several instances of damage to ecological systems in more scientific terms in an idea-packed essay, titled "A Biotic View of Land" (Leopold, 1939). This essay was originally presented as a plenary address (June 21, 1939) to a joint meeting of the Society of American Foresters and the Ecological Society of America, with the text published that same year in the *Journal of Forestry*. "Ecology is the new fusion point of the sciences," Leopold declared, implying it had eclipsed physics in that respect. Admitting that his own field, what he called "economic biology," had often concentrated on parts of nature, not processes, Leopold acknowledged the necessity for a more dynamic, process-oriented approach, and he in effect denounced chunking in the evaluation of natural processes and changes in them. He also rejects the idea of a "balance of nature" that, he said, helps make sense of short-term and local oscillations, but he feared that its use encourages some to assume a single, static balance. Becoming explicitly metaphorical, Leopold rejects the image of the grocer's scale, and prefers the image of a "biotic pyramid," a complex, many-layered system of processes driven by energy originating in the sun through photosynthesis and flowing upward to support layers of consumers in "a biota so complex, so conditioned by interwoven cooperations and competitions, that no man can say where utility begins and ends" (Leopold, 1939, p. 728). Exit chunking.

Would Leopold be happy with an economic valuation of the biota in terms of willingness-to-pay (WTP) for commodities derived from it? Would he act on a cost-benefit analysis (CBA) that has a line for "existence value," or "WTP for 20 percent improvement of views in a national park"? Of course he would not. He was insisting that scientific and evaluative choices be made within the flow of ecological change, describing multileveled dynamics and using metaphors to show cross-scale implications, and then interpreting these within the grand metaphor of the mountain.

Leopold was also pointing toward ecosystem management, adaptive management, and effective evaluation of impacts at all levels, whether economic, ecological, or geological. Clearly he believed these systems function best when left alone, though the system can oscillate and accommodate ecological changes, which, in turn, drive evolutionary changes. But here Leopold, despite his obvious preference for natural landscapes and undisturbed natural processes, nevertheless turned analytic, and he stated the problem of scale and sustainability in crystal-clear fashion. Leopold's eloquence in this scientific paper deserves its own voice. Stating his ambivalence about the balance of nature metaphor, he continues,

> This interdependence between the complex structure of land and its smooth functioning as an energy circuit is one of its basic attributes.
>
> When a change occurs in one part of the circuit, many other parts must adjust themselves to it. Change does not necessarily obstruct the flow of energy; evolution is a long series of self-induced changes, the net result of which has been probably to accelerate the flow; certainly to lengthen the circuit.
>
> Evolutionary changes, however, are usually slow and local. Man's invention of tools has enabled him to make changes of unprecedented violence, rapidity, and scope. (Leopold, 1939, p. 729)

For Leopold, the key, then, was to examine the "violence, rapidity, and scope" of actions and identify those likely to have longer-term and larger-scale impacts (Leopold, 1939, p. 729). To put the point another way, Leopold anticipated the insight, so well stated by O'Neill, Holland, and Light, that "human-induced change generally has a faster pace than ecological change, thus preventing the numerous and subtle ecological checks and balances from operating as they might" (O'Neill et al., 2008, p. 157).

Most of our worries about sustainability are generated by the mismatch between the scales on which we understand, act, and have an impact, and the much slower changes at the pace of ecology and evolution. Leopold recognized this and responded by showering the problem with metaphors: crane marshes where cranes "stand on the sodden pages of their own history," wolfless mountains fearing their deer, pyramids of energy flows . . . on and on the metaphors flowed from Leopold's fertile pen. He talked about advantages and disadvantages of various metaphors, and above all he made them explicit and central to his communication of scientific and management ideas. Leopold's method of evaluation anticipates heuristic 10, above.

Today, students of conservation biology are finding useful models for environmental management of cities by relying, in part, on such ecological metaphors as "ecosystem health," "ecosystem integrity," and "resilience" as they work to understand mixed natural and human communities (Pickett and Cadenasso, 2002; Kolasa and Pickett, 2005; Walker and Salt, 2006).

Having followed Leopold's reasoning—scientific, ethical, and metaphorical—with respect to the radical transition that was required of him when he learned that his wolf and deer management experiment was a failure, we can now extend that reasoning to yet another scale as a response to the current quandary. Economic growth and energy use have again expanded the scope of environmental problems, as the impacts of human activities affect the atmosphere that envelopes the earth, raising problems with effects that transcend even Leopold's expanded conception of a mountain and its ecosystem.

With this expansion, humanity now faces a third generation of environmental problems that demand attention because they are likely to affect every scale below them. As more and more humans produce and use more and more fossil fuel energy, the effects are changing larger systems that once set the constraints on smaller-scale systems and activities. Just as Leopold's overly violent removal of wolves created cross-scale spillovers that challenged his simpler view of game management activities by changing what he thought was constant, we humans today must recognize that the rapidity and violence of our activities are now spilling over into our global atmospheric systems. Just as starving deer were indicators that Leopold's actions on the mountain were too violent, today we endure hurricanes, floods, droughts, and extreme weather of all kinds, which provide similar indications of human-induced violence at the level of atmospheric systems. Leopold would perhaps offer two pieces of hard-learned advice: when undertaking or risking major structural changes in larger systems, first, act experimentally, creating limited controls and carefully observing changes to larger, normally slow-moving systems, and second, do not act irreversibly until you understand the likely consequences of your actions.

There is thus a structural analogy between the expansion of consciousness that humans now need and the expansion that enabled Leopold to self-critically reject one metaphor and the values it represented in favor of a more inclusive one—that of the "mountain." I and my research team, in our attempt to understand how environmental problems are bounded in space, began to think of Leopold's idea of "thinking like a mountain" as a sort of expandable simile that dramatically represents the importance of getting the scale right when one characterizes an environmental problem. It became natural for our members to say things like "We need to think like a forest" or "Let's think like a watershed" on encountering an example where fragmented management or irrational boundaries in jurisdictions frustrate progress. The simile seemed apt when we examined the epistemological and social processes involved in rethinking the management of bays and rivers in a more holistic manner. This simply involves transferring the simile from one ecological system to another. Managers of Chesapeake Bay's ecosystems expanded the scale and complexity of the mental model of bay functioning. Leopold's simile seemed to provide insight when applied to just about any second-generation problem.

Then, during a conversation with Paul Hirsch, my graduate student at the time and now my collaborator on many projects, I mentioned that I'd been asked to address a workshop-style conference sponsored by Defenders of Wildlife, on the topic of the fate of biodiversity in the age of climate change. My friend Rodger Schlickheisen, who was president of Defenders of Wildlife at the time, had asked me to provide an ethical point of view on the subject of threats to biodiversity from climate change. I knew the rest of the conference day would be full of bad news from top scientists who could show slides from about every ecosystem type in the world already displaying severe signs of stress from climate change. I told Hirsch I was struggling to come up with a clear and cogent way of addressing the problem of climate change and attendant threats to biodiversity. He said he had been talking with one of his classes about climate change and he'd made a PowerPoint presentation. Did I want to look at it? Sure, I replied; anything to help me come up with something clever to say about ethics in the face of biological calamity.

Hirsch's PowerPoint clicked with me as soon as I opened it: he had included the phrase "thinking like a planet" in several remarks and diagrams. I realized that, while Leopold is no longer alive, his simile is, for he left us with an insight that can apply to many situations. Paul Hirsch's fresh thinking, however, nailed down the huge point that was waiting to be made: that Leopold's scale-sensitive metaphors had catalyzed a remarkable transformation both in his own consciousness and also in the physical space and temporal scale at which he acted (Hirsch and Norton, 2012). He entered what Stuart Brand calls "the long present" (Brand, 2010). And this, in turn, had implications for the systems and the dynamics to which he paid attention. That is exactly what is necessary if humans are to successfully navigate through the many dilemmas of climate change and other third-generation problems. Since we humans now determine the future of the earth, it will be necessary for us to scale up Leopold's multiscalar insight of a mountain (ecosystem) to a multiscalar insight about the planet (atmosphere and all).

It will be vital to expand the scale of interest, modifying our conceptions of "home" to include a much larger area, and for our "present" to stretch to generations. In the remainder of this section, I explore what can be learned about climate change by extending Leopold's simile from ecosystems to our planet.

Notice, first of all, that Leopold was doing good science, counting and estimating deer herds; he was immersed in the data. Second, he engaged the data both as a scientist and as a manager, so he was thinking as an actor who had values and who viewed natural dynamics for both their beauty and their usefulness. His science was mission-oriented and embedded in his work for a major land management agency when he experimented with deer-herd enhancement by removing wolves. Third, it is above all important to see that Leopold's transformation, involving both a change in scientific understanding and a shift in the values he explicitly associated with those dynamics, was catalyzed and explained by a shift in metaphors from a productive factory metaphor to seeing land as a "fountain of energy." This new metaphor was both dynamic, in emphasizing energy flows and processes, yet it was also integrative. It does not replace the ecosystem-as-productive-factory metaphor so much as it relocates its productivity for human use within a larger system of food chains and energy flows. Fourth, even as Leopold was developing his new conceptualization of a large ecological system with some parts devoted to production and others to wilderness where wolves and the processes of which they are a part can continue, he joined with like-minded friends and cofounded The Wilderness Society. Identifying and focusing on the problem coincided with the development of an institution devoted to understanding and managing land on multiple scales and with differing purposes. The Wilderness Society, of course, has over the decades been joined by many other institutions and organizations in developing a strong voice in favor of protecting important biological processes.

Finally, Leopold's transformation—from utilitarian manager to an advocate of ecology-guided, multiscaled management that takes responsibility for all the impacts of management choices—involved value change. Wolves morphed from being "vermin" to guardians of the health and integrity of grazing ecosystems. This change included an important alteration in Leopold's science: he had to adjust his mental model and change assumptions about which dynamics are being engaged by his choices. All these dramatic changes were captured in a simile. This simile projected the scientific information he had gathered and the values he was acting on into a new context. Learning to "think like a mountain" was Leopold's escape from an inappropriate model as well as an overly simple dominant metaphor of nature as a productive system; the rest of his life and work became one of acting within the new, expanded context he had created.

Today, the discourse about climate change is fragmented, and most of the official discussions of policy use economic models, just as Leopold worked against a backdrop in which the resource agencies were entirely captured by a dominant metaphor of production and consumption. In the following sections, I explore the possibility that fresh metaphors can be developed that will reorient thinking, replace us humans within a global system undergoing change in "the long now," and reintegrate the processes that they depend on into a more complex system that will support the continuance of human society.

10.3. Ideas and Action: Callicott on Leopold on Planetary Ethics

Since the 1980s, I have emphasized the importance of Leopold's thinking like a mountain simile as a breakthrough idea, given its emphasis on the importance of getting the ecological scale right so as to understand and act appropriately in the natural world (Norton, 1990). Leopold's powerful and adaptable simile articulates how environmental managers and many others can make large mistakes affecting their environment when they develop small-scale mental models that encapsulate a problem in a system that ignores important variables (Walker and Salt, 2006, pp. 4–5). Leopold learned to think like a mountain when he realized that his actions to increase the deer herd in the Southwest would disrupt larger-scale and normally slower-changing systems—those ecological and evolutionary processes that protect the "mountain" at larger scales. His intention to create a hunters' paradise set in motion events that irreversibly damaged natural systems because he paid attention to impacts of his wolf elimination only on the scale of hunters, deer, and wolves, ignoring the larger-scale impacts on the ecological and evolutionary systems on the mountain.

So it was natural for Paul Hirsch and me to once again mine Leopold's versatile and insightful simile. J. Baird Callicott has recently adopted a similar strategy, surveying Leopold's thought for insights that will help us to understand our obligations in a world of anthropogenic climate change (Callicott, 2014). Since Callicott, despite developing many of the same ideas Hirsch and I published in 2012, presents his argument by contrasting it with some of my earlier ideas, I devote this section to clarifying where Callicott and I agree and where we disagree when it comes to learning from Leopold regarding how to respond to climate change. Here I will, first, concentrate on the ways Callicott and I have converged on similar Leopoldian lessons applicable to climate change, and then will go on to show how Callicott's approach to climate change is limited in important ways.

On one key point, Callicott and I agree: addressing climate change—a third-generation problem—will require scale sensitivity (Norton, 1990). I take the argument of his essay "Thinking like a Mountain" (Leopold, 1949) to imply a kind of multiscaled pluralism: humans value nature in multiple, sometimes incommensurate ways, depending on the system scale on which they are focused. Callicott, in *Thinking like a Planet*, asserts that "the land ethic . . . is ecologically scaled, both temporally and spatially" (2014, p. 149). And he continues, "The land ethic, in short, is inappropriately scaled to meet the challenge posed by global climate change" (p. 150) and climate change "brings forth new concerns for which the land ethic has no conceptual resources" (pp. 150–51).

I have discussed many of the differences between Callicott and me regarding the details of Leopold's thought and its implications elsewhere (Norton, 2013),[1] but our major differences regarding what should be learned from Leopold regarding climate change can be encapsulated in three important disagreements, which I will develop in turn.

1. Callicott thinks the distinction between anthropocentric and nonanthropocentric foundations is of great importance in understanding and getting environmental ethics right. I, on the other hand, think this distinction, while interesting as an abstract question in philosophy, is determinative of little or nothing of importance in environmental policy.

2. Callicott embraces a form of pluralism with respect to "conceptual resources" available to discuss environmental problems emerging on different scales. He thinks these differing ethical foundations can all be reconciled by appeal to his overarching ethical theory, which he attributes to Hume and Darwin as well as to contemporary thinkers who pursue a "science of ethics." I, by contrast, while agreeing on the importance of naturalistic ethics, think the moral quandaries and recommendations at different levels are, and will indefinitely remain, irresoluble on purely intellectual and moral grounds. Ultimately, decisions to favor the considerations of one scale of problems over those of another will have to be political, never intellectual.

3. Growing out of disagreement 2, we also disagree regarding how human behavior can be changed. Callicott believes that developing, advocating, and enforcing a science of ethics will lead to a transformation in public attitudes based on better ideas of ethics. I believe, on the contrary, that

better decisions will result from better decision-making processes and that behavioral change is more likely to result from changes in the incentive structures under which people make decisions than on simple changes in attitude rising out of abstract ethical reasoning.

10.3.1. Anthropocentrism and Scale in Environmental Thought

As he has done on several occasions before, Callicott separates three forms of anthropocentrism—metaphysical, moral, and tautological—as an entrée into his discussion and comparison of his version of the land ethic with his introduction of his earth ethic (what he calls his ethic for climate change). Callicott understands metaphysical anthropocentrism, such as that of Aristotle, as the belief that "human beings occupy a privileged place in the order of being" (2014, p. 9). Moral anthropocentrism and its denial, moral nonanthropocentrism, can be, but need not be, based on their metaphysical counterparts, and Callicott bases his nonanthropocentrism on his Humean ethics rather than on metaphysics. Callicott also articulates what he calls "tautological anthropocentrism," which he describes as the principle that "All human experience, including all the ways that human beings experience value, is human experience and therefore tautologically anthropocentric. Tautological anthropocentrism is humanly inescapable." It follows, he thinks, that tautological anthropocentrism comes only to the assertion that "all human valuing is human valuing," a triviality from which nothing of interest can be inferred (p. 10).

It should, however, be noted that Callicott has, on numerous occasions, also insisted that all values are necessarily human values—they are attributed to nature by us humans, rather than existing independently (Callicott, 1992). Combining these two "trivialities"—that all human values are human values and that all values are human values—it is possible to derive a fourth form of anthropocentrism. For lack of a better term, I refer to this form as "administrative anthropocentrism," which is the recognition that, when there is a conflict among values held by persons or groups, the resolution of such conflicts, if achieved, will be a resolution by human beings. Thus, we note that in some cases, necessary truths, while trivial in the sense that they assert no empirically falsifiable claims, can constrain belief by ruling out certain possibilities such as, in this case, the possibility of resolving human value conflicts nonanthropocentrically. This limitation on conflict-resolving discourse, while based on definitional and hence necessary truths, means that anthropocentric arguments will have to play a key role whenever environmental and other human values require reconciliation.

10.3.2. Resolving Cross-Scale Conflicts among Scaled Values

To evaluate Callicott's proposed "earth ethic" governing actions that drive climate change, one must address his new emphasis on the pluralism of Leopold, since Callicott now follows me in recognizing that protecting the environment requires appeal to values that emerge at different scales. Since his first publications in environmental ethics, Callicott has faced a significant dilemma: he believes that Leopold rejected anthropocentrism in favor of "holistic nonanthropocentrism." As pointed out by Tom Regan, without qualification, a holistic ethic that prioritizes ecosystems and species over individuals will apparently conflict with our deeply held beliefs about obligations to individuals (Regan, 1983). Strongly charging Callicott (and Leopold as understood by Callicott) with "environmental fascism," Regan showed that an unqualified moral preference for whole ecosystems and species over obligations to individuals (much as Germany's National Socialists valued the state more than individuals) would undermine the most basic ideas of ethics.

Callicott, while emphasizing that the land ethic is holistic "with a vengeance"(1989, p. 84), nevertheless rejected Regan's charge. His defense is based on Leopold's explanation of the expansion of ethics since the days of Ulysses as one of expanding circles of moral concern, beginning with family, expanding to include one's clan, members of one's country, and eventually all humans. Leopold's recognition that our culture does not currently have an ethic regarding the land, and therefore we treat land as property, leads him to suggest that a land ethic is one more accretion to these concentric circles. An ethic, according to Leopold, "is a limitation on freedom of action in the struggle for existence" (1949, p. 202). Despite the apparent cogency of Regan's charge, Callicott responds, "The land ethic, happily, implies neither inhumane nor inhuman consequences." His explanation is that "[t]he biosocial development of morality does not grow in extent like an expanding balloon, leaving no trace of its previous boundaries, so much as like the circumference of a tree. Each emergent, and larger, social unit is layered over the more primitive and intimate, ones." Explaining the layering effect, Callicott writes, "Family obligations in general come before nationalistic duties and humanitarian obligations in general come before environmental duties. The land ethic, therefore, is not draconian or fascist. It does not cancel human morality. The land ethic may, however, as with any new accretion, demand choices which affect, in turn, the demands of the more interior social-ethical circles" (Callicott, 1989, pp. 93–94).

So Callicott avoids the charge of fascism by prioritizing earlier accretions over later ones. But here the dilemma arises: Does Callicott mean to say that, fairly automatically, when there is a conflict between obligations to humans and the land ethic, the land ethic always loses to the "more primitive and intimate ones"? Or does he mean to say that, when human and ecological values conflict, a negotiation must occur in which either value might prevail but where other considerations—such as the egregiousness of threatened ecological damage—might override the priority of the ethics for the inner circles? In other words, is the priority of familial and other anthropocentric values dominant? If so, Callicott's nonanthropocentrism (and Leopold's on Callicott's interpretation) apparently collapses into a less blatant, but no less systematic, anthropocentrism. If, on the other hand, he significantly limits human behavior in the face of possible ecological damage, then he owes us a means to decide exactly when the land ethic overrides obligations to humans and when the priority of inner levels protects humans. Callicott has never decisively resolved this dilemma, and I turn now to show how it once again arises as he layers on a new accretion: his earth ethic.

Since there would be no point in adding an earth ethic unless it somehow limits behaviors based on prior layers in the system, I assume Callicott believes that there will be some cases in which the earth ethic, despite being a later accretion, will question and perhaps override either obligations deriving from the land ethic or prior ethical obligations for dealing with humans. Having finally recognized scale-sensitivity in ethics, Callicott sets out to reconcile the land ethic with the earth ethic by applying his rule of layered obligations. He then argues that the earth ethic and the land ethic (as well as standard human ethics) can be reconciled by showing that both ethics appeal to the same underlying Humean moral theory. According to this approach to ethics, ethical behavior is genetically rooted in the emotions: humans respond to situations that affect their feelings. The question that dogs Callicott is whether a purely emotive theory of ethics can adjudicate moral quandaries across scales. If not, how could an affect-based ethic determine which scale is appropriate to consider a given quandary? If so, we need some explicit guidance, which I cannot find in *Thinking like a Planet*.

I now turn to a discussion of this theoretical reconciliation of the earth ethic with the land ethic. Callicott recognizes that explicit representation of multiple spatial and temporal scales can affect ethical reasoning, and that adding an earth ethic to the land ethic requires some kind of reconciliation: "The temporal scales of the Leopold land ethic, however, are incommensurate with the spatial scale and temporal scales of global climate change" (2014, p. 288). So Callicott ends up with a hybrid system of ethical rules that, at the scale of human families and societies, is at once anthropocentric (at the scale of individual human ethics), nonanthropocentric (at the scale of ecological, or "land," ethics), and then (at the even larger scale at which climate change threatens human civilization) anthropocentric once more.

He cashes out concern for the abstraction of "indeterminate distant future generations considered holistically or collectively," by offering a "palpable surrogate": *global civilization*, which he describes as "palpable, precious and fragile" (2014, p. 302). He thus introduces three ethics, each governing at its appropriate scale. He embraces as one of the "principal philosophical tasks" of his book on planetary ethics "to unify, theoretically, our familiar human-oriented ethics, the land ethic, and also the Earth ethic—to articulate a theory of ethics . . . that will embrace them all" (p. 11). He says, "In a single system of moral philosophy, which I attempt to achieve in this book, ontological pluralism and deontological pluralism must be tempered by theoretical monism" (p. 12).

Callicott thus recognizes that different moral criteria will be appropriate at different scales: "Most of our daily moral quandaries do not involve a choice between doing the right or doing the wrong: rather, most involve a choice between fulfilling duties generated by membership in one social order and fulfilling those generated by membership in another." (p. 300). So ethical actors should respond to different emotive affects as their concern shifts from one scale to another. In some cases, it will be appropriate if a person's concern for family overrides concern for the land, and apparently, it will sometimes be more appropriate to act on emotions involving threats to human civilization itself rather than to concerns about individuals and ecosystems.

For example, if one makes a decision based on the individual human scale when managing an ecological system, as Leopold did when removing wolves, then mistakes will be made. Since his mental model of the problem was not enough deer, he used a simple deer-wolf-hunter model. This meant he was excluding impacts of wolf removals on the larger ecosystem—the mountain. Similarly, if one were to protect an ecosystem, ignoring the effects on the atmosphere at the cost of human civilization, one would have erred: "The land ethic is, therefore, a poor fit with the most urgent and dire environmental concern of our time. To have some chance of confronting global climate change successfully, we need to be equipped with an environmental ethic that is commensurate with its spatio-temporal scale" (Callicott, 2014, p. 300).

Will the Hume–Darwin ethical tradition serve to unify his spatially disparate ethics? Can appeal to human feelings and emotions determine which ethic is to be dominant in a given situation? I agree with Callicott that introduction of scalar analysis is essential to understand and respond to complex threats to the environment. Consider a hypothetical but plausible case. Suppose it has been determined that a particular ecosystem—say, a wetland—

has been degraded by human activities and there is a debate about whether and how to restore the system. Suppose also that the only means available will require significant use of fossil fuels for dredging to remove toxins and other matter. On one side, ecologists argue for the importance of restoration because of impacts on wildlife and adjacent systems; atmospheric scientists argue, however, that the restoration will contribute significantly to climate change, which has become so dire as to threaten future human civilization. Can Callicott, armed with Humean ethics, resolve this disagreement among well-meaning environmentalists? I doubt it. Nothing in that emotivist tradition in ethics can determine which strongly felt emotions should trump which others (Stevenson, 1937).

We can now summarize where this argument has carried us: Callicott, by introducing scalar incommensurabilities in ethics, exacerbates his long-standing dilemma: how can ethical human beings decide an ethical quandary that apparently pits a legitimate emotive concern for humans against a legitimate concern for ecosystems and even for human civilization? In my view, Callicott will never succeed in resolving this dilemma because, in conflicts such as these, mere intellectual or conceptual resources can never succeed. For this reason, I have developed the themes of this book around how to improve democratic processes so that conflicts such as these can be resolved through negotiation, institution-building, and compromise rather than by appeal to theory.

10.3.3. Attitudes and Decisions: Cognition and Action

My skepticism regarding Callicott's proposed theoretical reconciliation of obligations appropriate at different scales highlights my third disagreement with Callicott's reconciliation of the three scales included in his hybrid emotivist ethic: we disagree about what is likely to change human behavior and, consequently, we disagree regarding the strategy we should pursue if we wish to save the world. Callicott has articulated what he calls his "agenda for future environmental philosophy," as follows: "First, we identify and criticize our inherited beliefs about the nature of nature, human nature, and the relationship between the two. . . . Second, we try to articulate a new natural philosophy and moral philosophy distilled from contemporary science. We try, in other words to articulate an evolutionary-ecological worldview and an associated environmental ethic" (Callicott, 1995, p. 41). The contribution of environmental ethicists, he thinks, stands at the top of the list of important forces driving the environmental movement (p. 42), and this results from the fact that "In thinking, talking and writing about environmental ethics [he and his nonanthropocentric colleagues] already have their shoulders to the wheel, helping to reconfigure the prevailing cultural worldview and thus helping to push general practice in the direction of environmental responsibility" (p. 43). Clearly, Callicott believes that changing cultural attitudes is the primary means by which the environment must be saved, and that philosophical analysis of ideas will be the driving force carrying society into a new millennium of moral rectitude regarding the environment.

What I find interesting is that Callicott, while clearly well versed in, and deeply committed to, an important role for cognitive psychology in understanding ethics and ethical responsibility, he seems entirely oblivious to equally compelling scientific developments relating to cognition and action. As I note above, in section 5.4.3, Thomas Heberlein (2012) has recently summarized the findings of psychologists who have studied the relationship between attitudes and actions. We derived from Heberlein's work the following apparently solid conclusions:

1. Attitudes change, but only very slowly and only under the right conditions.
2. Attitudes, by themselves, are usually not sufficient to lead to appropriate behaviors and, as a corollary, changes in attitude are unlikely to change behavior.
3. Efforts to change behavior are more likely to succeed when the desired behavior is consistent with existing attitudes and more difficult if inconsistent with such attitudes.
4. "Cognitive fixes," the provision of information and attempts to change behavior through education alone, almost always fail.
5. Behavioral change is much more likely to occur as a result of a "structural fix": "a change in the social environment that influences what people do" (Heberlein, 2012, p. 6).
6. Structural fixes, such as changing incentive structures, are much more likely to result in behavioral change; accumulating behavioral change can lead to the emergence of new norms.
7. Encouragement of new norms as a result of structural fixes is more likely than other strategies to result in behaviors productive of environmental protection.

These conclusions, based on the psychological study of cognition, conation, and behavior directly challenge Callicott's most central assumption: that better environmental policy will result from academic and educative attempts to change people's beliefs and attitudes and that, if the world is to be saved, he and his nonanthropocentric colleagues will certainly be the leading contributors to that process.

Thus, even if Callicott were to succeed in answering my arguments in subsection 10.3.2 by offering a hybrid system of ethical thought that contained within it the resources to determine in each case whether we should apply human ethics, the land ethic, or the earth ethic, his approach will still fail because it locks into a strategy that has been shown, based on a multitude of empirical research on attitudinal and behavioral change, to be unworkable.

Callicott's approach to changing behaviors resulting in climate change can be contrasted with the approach in this book. I do not believe that, when important conflicts occur—such as differences regarding whether a wetland should be restored, even if the fossil fuels used in the restoration contribute strongly to climate change—they can be resolved on the level of competing attitudes, ideologies, or monistic theories of ethics. Consequently, I have emphasized procedural rationality supplemented with empirical work on which processes actually work and which processes are likely to fail.

Heberlein's work is a paradigmatic case of this kind of empirical work: his conclusions imply that real progress in environmental protection will occur when the public is engaged in an adaptive process that is rich enough to include direct engagement with policy issues within concrete, particular problem contexts. Such engagement can and should include deliberation, some of which will be of the abstract variety that Callicott favors; yet this abstract and ideological reasoning will not lead to better decisions by itself. This reasoning must be embedded within a broader adaptive process with the natural sciences, social sciences, and humanities. Even more important, the process must not merely present ideas and arguments; it must include also proposed actions, as, for example, "We should set up a monitoring system," or "we should provide tax breaks for certain behaviors," or "we should engage citizen juries and employ all the exercises and tools discussed in chapter 7."

The priority of process does not end ethical discussion and debate, but it does embed that discussion in particular problem situations. Heberlein's findings drive this point home: it is usually structural factors in particular situations that govern behavior, so if one wishes to encourage behavioral norms that protect the environment, one should pay more attention to local situations and less to abstract, monistic theory-building. Adaptive management processes, far more than abstract monistic thought, is more likely to lead to agreements including structural changes that participants agree will cause behavior, including their own, to alter and encourage the emergence of new behavioral norms.

10.4. The Special Challenges of Rapid Climate Change

While I am convinced that we can, and must, learn from Leopold, we should also be realistic. The situation faced today differs in many ways from the problems he faced and reformulated in more holistic terms. First, Leopold was an extraordinary individual, who combined technical knowledge, unquenchable curiosity, unusual powers of observation, and a philosophical bent that apparently had no place for ideology or dogmatism. Recognizing unusual Leopoldian characteristics, those who intend to follow in his path must accept a huge challenge: can they approach emergent global problems with the openness to learning that drove Leopold?

Second, because Leopold provided such a graphic, personal, and convincing account of his own transformation, it is tempting to simply say, today, that seven billion similar transformations must take place on an individual basis. Leopold, as noted, did not just change his own mind—he set out to change the minds of others through the launching of The Wilderness Society, and he was far ahead of his time in thinking of the employees of resource agencies as exercising primarily a teaching and collaborative function. Is it realistic, then, to hope we will all be able to follow Leopold's lead as we face third-generation environmental problems?

Third, Leopold's ideas have penetrated multiple fields and have led to new organizations, and as noted, his ideas are now embodied in the movement toward adaptive management. Indeed, Leopold has probably had greater impact on the American culture of conservation than any other historical figure; so his contribution has become social and cultural through the gradual processes of teaching students, affecting organizations, and so forth. Today, the Aldo Leopold Foundation spreads Leopold's ideas, even as it continues to refine his conservation practices. Yet this shift has been a long time in coming; indeed, it is still far from complete. A transformation in cultural models regarding climate change and the atmosphere—learning to think like a planet—must apparently take place extremely rapidly over the whole world. The pace at which Leopold could transmit his ideas, converting a few colleagues and a few thousand readers with his books each year, will not be adequate today.

If the challenge faced today is to duplicate Leopold's metaphorical transition from old-think to a new vision for ecosystem management in response to climate change, then this further expansion of models adds significant complications (Gardiner, 2011). First, the change cannot simply propagate on an individual-by-individual scale. However successful Leopold's ideas of adaptive ecosystem management have been in convincing managers and environmentalists of the importance of scaling up to show concern for larger ecological systems, the change that will be required of the earth's people can only be successful if it includes a collective shift in understanding and consciousness to a scale much larger than that observed over thirty years in the Chesapeake Bay region, for one significant example. Second, Leopold's message, while emphasizing the need to understand resource use as embedded within larger ecosystems, was directed at his fellow Americans and colleagues in the US resource agencies. Success in adaptively managing through climate change must be capable of addressing the problem at a global scale, and the message must reach a large portion of the world's population, creating a shared sense of responsibility to arrest or at least ameliorate earth-scale changes.

The urgency of the message regarding climate change, and the vast human populations that must all, or almost all, make a crucial transition from old-think to new-think, will pose unique challenges. There are, however, two reasons for optimism.

The first reason is a truism—the challenge of growing interrelatedness of all scales in the system is matched by acceleration in expansion of communication. Today, the new, connected world of electronic communication has provided many examples of events and ideas "going viral," as websites pick up ideas and as media cover the emergence of the phenomenon, which, in turn, leads to more traffic and more media coverage and explosion into worldwide social media. While such events are difficult to predict and even more difficult to instigate, their existence means that, once important new thinking about climate change begins to occur, those ideas can be transmitted almost instantaneously throughout the globe. This does not mean that adaptive managers can cause a major change in mental and cultural models on demand; but it does mean that, once a planetary consciousness emerges, the possibilities of its spreading are greatly enhanced. For these reasons, it may not be overly optimistic to think that, by engaging aggressively in experiments in group social learning, a major breakthrough is possible in the cultural models associated with large-scale climate systems.

The second reason for optimism is the advance of social sciences, especially the cognitive sciences, which has greatly increased understanding of the mental and social processes by which people learn. These processes could trigger public participation in a mass transformation of human understanding through public participation and improved decision making.

While this concept of collective cultural change may seem mysterious, there is actually a relevant and very helpful interdisciplinary literature in cognitive science, reinforced by work in the philosophy of science, which explores the concept of "distributed cognition" (Hutchins and Klausen, 1996; Giere and Moffatt, 2003; Nersessian, 2006). According to this research, collective learning can take place through external representations, objects occurring outside individual minds, which can be essential elements in thinking processes. Most persons can do simple math problems in their heads, but as soon as aids to calculation are added—from use of fingers to use of an abacus to use of a computer—the capacity to solve math problems increases exponentially. These objects and tools facilitate the thinking process and make it more public and communicative, capable of engaging more individuals' experience. Understanding how such collective changes might take place requires taking into account the entire cognitive system of thinking individuals.

Distributed cognition is essential for many complex tasks, such as landing a jumbo jet. No individual can land a large jet; the pilot relies not only on the copilot and crew but also on complex instrumentation, communications from air traffic control, and so forth. All the skills and all the information associated with landing an airplane constitute a complex process involving distributed cognition. This may include parts of a person's body, written symbols, instruments, and all the relations among them.

Similarly, in specific situations where community managers are trying to act adaptively, they can rely on shared knowledge that will support the cooperative development of a shared model of the dynamics relevant to important management decisions. While the creation of a new consciousness and a new metaphor for understanding the processes of climate change are much more complex even than landing an airplane, this example illustrates the advantages of addressing complex problems more distributively. An expanded model of cognition, a model that introduces many additional tools—including metaphors, climate models, images (of melting glaciers, for example), and actual occurrences—will contribute to the collective thinking that must be engaged in if the earth's inhabitants are going to learn to think like a planet. Such emergent models could become effective boundary objects that will be instrumental in catalyzing cultural change through careful attention to metaphors and to integrative models that help the public to develop new understandings of processes affecting everyday events like storms and droughts. The opportunities for adaptive managers to develop clear and compelling boundary objects may prove to be a powerful tool of social and cultural learning, especially if they encourage social learning regarding distributed knowledge.

For example, it may turn out that the further development of climate change models will result in the development and propagation of a much simpler model than the technical ones now being built. Such a model, while probably lacking in many details, might be reasonably consistent with the technical models, but should be simple and clear enough that scientists from multiple fields as well as members of the public can understand and rally around it. Ideally, such a model would be consistent with and able to incorporate detailed scientific knowledge and also to evoke important, widely resonant metaphors.

Perhaps the closest anyone has come to suggesting a metaphor that could be transformative with regard to climate change discourse is James Lovelock's invocation of the ancient earth goddess Gaia, which has an almost-spiritual aspect functioning in tandem with respectable science about positive and negative feedbacks arrayed within sophisticated hierarchical models of nature (Lovelock, 1989). In this respect, it is similar to Leopold's simile of thinking like a mountain: it mixes scientific theory with evocative, metaphorical, and highly evaluative language associated with a spiritual and yet scientifically driven earth goddess force that acts according to broadly general principles that can be insightfully understood as involving agency and, at the same time, evoking timeless values.

Box 10.1. James Lovelock: Independent Scientist, Inventor, and Gaian Theorist

James Lovelock (born in 1919) struggled financially to get an education, studying several subjects opportunistically, and eventually falling into an opportunity to gain a PhD in medicine. Later, he credited this spotty career in education as shielding him from becoming an overspecialized and uncreative researcher. He became an independent scholar and inventor and is credited with important inventions.

Working on assignment for the US National Aeronautics and Space Administration (NASA), he was enlisted to develop a device that might detect life on Mars. Lovelock, short-circuiting the effort, developed a theory that, on earth, life contributes to the conditions (far from equilibrium) that are conducive to life and, therefore, Mars (the atmosphere of which is in equilibrium) could not, currently at least, support life as we know it.

Building on the insight that a system cannot be maintained far from equilibrium by accident, he formed the hypothesis that life, through a complex system of feedbacks, regulates the macro-systems within which life develops. At first, scientists too influenced by positivist constraints on scientific theory ridiculed Lovelock's hypothesis for violating any known causal forces and for suggesting teleological forces active in nature. Further alienating hard-nosed scientists, Lovelock called his speculation the "Gaia hypothesis," after the early Greek earth goddess, which was thought to violate strictures against mixing metaphor and religion into scientific study.

Lovelock, always the renegade, has adopted positions orthogonal to accepted opinion as he recognizes the role of life in all its diversity as the driver of creativity. His Gaian metaphor, open to interpretation and reinterpretation, stands as an example of how scientific understanding can evoke poetic flights of fancy.

Some unfortunate early attacks on Lovelock's theories by scientists, too enamored of positivism and unfortunately not in tune with the creative dialectic between metaphor and measurement, deprived the Gaia metaphor of some of its persuasiveness. But today, the scientific analogues of Gaian theory do help unify understanding of change over multiple and embedded scales of time and space. Since so much intelligent conversation has been provoked by Lovelock's bold strokes of theory—including some highly cogent criticisms of it—I think it unnecessary to go into detail here, but ask a question: Recognizing the need for a more expansive and complex metaphor as a helpful guide to making sense of the scale at which climate change takes place, can Gaian theory, perhaps modified in some ways, become a transformative metaphor for learning to think like a planet? One observation that points up a huge challenge to following Leopold's example in macroscoping to a global scale in addressing climate change is that, despite considerable science and attempts to communicate the gravity of the situation to the public, the process of building successful, effective, and adaptive global-scale institutions lags well behind scientific progress. Time will tell whether the process outlined in chapter 5, the emergence of a "public," can create a force in an international public sphere. Will a true, Deweyan public arise on a global scale, gain a voice in the public sphere, and then congeal into a world-scale community that can, through collective learning and bottom-up organization, generate a public discourse in which information is presented in digestible ways and old assumptions and inappropriate metaphors can be reconsidered and replaced? One should not minimize the magnitude of the task. One thing is clear: if Leopold were alive today, he'd already be acting, already be designing his actions as learning experiences, and already working to develop appropriate global-scaled organizations and institutions that could deliver us from the current state of confusion and place us on a path toward effective, world-scaled action.

Before ending this discussion of the possibilities and barriers to creating an expanded model of the environmental challenges in the third generation—this global age of environmentalism—a major conundrum implicit in the line of reasoning developed in this book must be addressed. On the one hand, I have tried to capture and support a significant trend in conservation thought that, given the nonrepeatability of complex problems, it is necessary to emphasize the uniqueness of local situations facing environmental quandaries. On the other hand, while I have tried to discuss this growing consensus toward deliberation and local involvement, I have also been acutely aware that the emergence of third-generation problems, especially climate change, thrusts global issues and problems onto the center stage of environmental policy consideration.

Following arguments that emphasize processes, especially local ones that are necessary for progress on wicked problems, a locally based approach to adaptive management has been developed that incorporates science, evaluation of change, and public deliberation in an integrated discussion of local problems. While I have supported the expanding view that effective adaptive environmental action must be particular and local, this consensus may seem to conflict head-on with the other important trend that also inspires this book—recognition that, today, third-generation problems that encompass the whole earth and its atmosphere must be addressed. The pace of environmental change will, for the foreseeable future, be driven by global forces. Can this conflict be reconciled? Can climate change be addressed by developing policy mainly from the bottom up?

I believe the answer to these questions can be positive. Despite serious constraints on local activities from power structures at larger scales, local communities can catalyze action if they are organized and pursue effective processes. Local and regional groups and agencies are currently in the forefront of experiments to reduce emissions in their own communities. A reconciliation can be achieved by noting that localism and globalism actually amount to two perspectives on the same processes. As was learned from hierarchy theory (HT), it is useful to think of physical systems as nested hierarchies in which each component is both a "part" of the larger system and also a semiautonomous whole. Thus, observation of a multilayered complex system involves two perspectives: one can either look at the system from the smaller end at components that can explain the mechanisms occurring at larger levels, or one can look at the larger, global systems as composites of multiple layers of smaller subsystems that change at a slower pace that creates constraint and opportunities at smaller scales. Whichever perspective we take as primary, the other perspective is implicated, even if unexpressed.

Environmental problems can thus be thought of as occurrences in which some element or subsystem begins to change and develop in novel ways. This organization implies that a correct analysis of global problems sees them as a manifestation of effects of rapidly and violently changing smaller subsystems within the global system. Looking at the problems from the local perspective, then, many actions that will affect global problems for good or ill will show up as local imbalances and unsustainable forms of local development. I think current international efforts to address climate change are extremely important, for without them it is difficult to see how to place—and implement—an international price on carbon, for example. But I'm placing my bets on local and state-level efforts to affect change. The excess carbon in the atmosphere is all coming from somewhere; and local and state experiments can form the core of an expansive adaptive management process that builds in multilayered efforts to mitigate climate change on all levels.

When environmental problems are seen from a local perspective, it is essential that this local perspective not be parochial, whatever scale one chooses to think of as one's "place." One's home place is embedded in larger places, and there will be spillovers across boundaries, both upward and downward through socioecological systems and subsystems. This idea is well captured by the geographer Yi-Fu Tuan, who said that, to understand one's place, one must also understand the space around that place (Tuan, 1971). One cannot escape acknowledging, in the age of climate change, that every country and every culture is one's neighbor.

So, in the normal course of events, conflicts across community lines can be expected, as complaints about spillovers across boundaries lead to tension. What should be the adaptive manager's response to these conflicts? It should be to focus on the larger problem, set out to create or stimulate a "public" of concerned citizens, and begin to work toward creating community on an expanded scale—one that is appropriate for the emergent problem or mess that must be addressed. Thus, the basic idea of developing an adaptive management process must be seen as evolving a multiscalar—or "polycentric"—system of institutions that eventually must span all layers, from local to global (Ostrom, 2007).

In this chapter, I have struggled with global climate change, recognizing that its effects will no doubt stress all levels of the systems of planet earth. If the accounts in the previous chapters are correct, it is possible to understand climate change as an emergent problem, one that still requires the development of a strong "public" able to affect important decisions on a global scale. At present, the likelihood that global institutions capable of motivating major changes in worldwide policies will be developed seems, at best, distant. This is a depressing conclusion, perhaps, given the acceleration of global change. Will an adaptive approach to managing earth's atmosphere, based in strong international institutions and supported by an increasingly articulate "public" that demands action at larger and larger scales, develop quickly enough to save the planet from catastrophic change that will outstrip humans' ability to adapt? I cannot answer that question. But I do claim that I have developed a general strategy that has a better chance than any other approach that has been proposed.

As desirable as it would be to jump-start the process of institution-building and international consensus-building that would be necessary to engage the whole world in a disciplined approach to addressing climate change, I have in this book favored the local-outward approach. Countless local adaptations may be the only viable way to address climate change issues experimentally, as societies struggle to develop world-sized public and effective global institutions to mount a disciplined global response. All the excess CO_2 in the atmosphere comes from some local place; therefore, the effects of climate change will be felt in each and every local place on earth. At present, the most effective means to address global climate change is to adapt local activities so that they can impact global dynamics in a positive way. For example, each community, if organized adaptively, will no doubt address its energy issues by exploiting locally available resources—solar in the desert, wind on the plains, and rescued gas from landfills in cities, for example—and diversity of institutions and experiments will increase the likelihood of social learning. Adaptive management—inspired by Leopold's guidance toward experimental actions and toward new mental and cultural models—may thus provide the only feasible response to runaway climate change.

Policy Analysis or Problem Solving?

Myths matter. In environmental policy discourse—especially among "experts"—a myth has it that, with more data and better analysis, the best solution to environmental problems will be found. This myth of the one best solution encourages policy analysis of environmental problems as an intellectual enterprise separate from, and prior to, management of ecological and mixed human-ecological systems. The idea of an intellectual analysis guided by a disciplinary theory suggests that, if one is simply patient, scientists will fill in the missing details and, voilà!, the analyst will provide theoretically grounded data for guidance toward improved and sustainable policy. This faith in the existence of a "right" answer or optimal solution, in turn, colors our understanding of environmental policy discourse. By constantly applying the in principle patch (IPP) and by assuming that all that is needed is more data, the unavoidable conclusion that no single correct answer exists is obscured and action is delayed. If one assumes that there is a single correct answer, then one tends to advocate policies that seem on the way to the right answer. Even more damaging, this myth reduces deliberation and management to manipulating and cajoling the public and participants to accept the "right" answer. It is time to forsake the IPP and recognize that it will be necessary to learn how to act while applying many partial models.

I teach in a school of public policy where many of my colleagues are excellent at policy analysis; I respect them for it. But sometimes—exaggerating to make a point—I say to my students, "I don't want to teach you to be only good policy analysts! Seeing public policy practice as one of analyzing and comparing proposed solutions implicitly presupposes that the best solution is on the docket. I don't want analysis! I want creativity! Never assume that the set of proposed solutions under discussion includes the best solution. Rather than coming up with a best solution that resembles a mythical optimal outcome, emphasize finding better solutions, given the situation, after a systems analysis of the problem."

Indeed, creative problem solving should be the focus of adaptive management. Evolution and adaptation depends on there being a diversity of actors who try different strategies—different adaptations. Adaptive management is problem oriented and thrives when many options are tried and experimented with, just as a species, struggling to survive in a changing environment, will do better if its members exhibit greater plasticity in behaviors.

As a closing gambit, let me develop this idea of creative response to problems and contrast it with theory-driven analysis by returning to the discussion of Robert Solow and weak sustainability, contrasting his views with the problem-oriented, multiscalar approach advocated in this book. As noted in chapter 4, Solow argues that, since the current generation cannot know what people in the future will want, sustainability should simply ensure that each generation pass forward at least as much wealth as was bequeathed to them.

In his paper titled "Sustainability: An Economist's Perspective," Solow poses the following "paradox" for the present-day devotee of sustainability: "[Y]ou should think about it as a matter of equity, as a matter of distributional equity, as a matter of choice of how productive capacity should be shared between us and them, them being the future." But then, he argues, "[o]nce you think about it that way you are almost forced logically to think about equity not between periods of time but equity right now" (Solow, 1993, p. 185). "The paradox arises," he says, "because if you are concerned about people who are currently poor, it will turn out that your concern for them will translate into an increase in current consumption, not into an increase in investment. To invest in the future today, while people remain hungry, is to shortchange the poor, so thinking about poor people today will be disadvantageous to the point of view of sustainability." Unpacking Solow's "paradox" will illuminate a more fruitful path forward.

Notice that Solow defines sustainability as simply a matter of maintaining undifferentiated capital: wealth. Note, too, that he judges investments according to the size of the investments, not their quality, that should be made for the future. More specifically, he does not distinguish between investments that increase options and protect opportunities for the future and those that simply generate larger revenues in the short run. Weak sustainability theorists, similar to Leopold in his misadventure with wolves, are in danger of making decisions without paying enough attention to impacts on normally slower-changing systems and to consequent impacts on future resource users. Hidden in Solow's argument that investments for sustainability will damage the current poor is a suggestion that any payoff on sustainability investments lies far in the future. If one makes investments in the present that improve the situation of the poor today, and if they are designed to be lasting and sustainable over generations, then the sharp distinction between concern for today's poor and tomorrow's possible poor disappears. What we need are creative, smart investments that move people out of poverty even as they protect the natural and social infrastructure to support both immediate improvements and sustainable progress. Investments that improve the lives of present persons, and that sustain these gains to the benefit of their children and their children's children, are the true path to sustainability.

A multiscalar understanding of resource and ecological systems aids in the search for creative solutions. If one thinks of impacts of human activities as occurring on multiple scales, and if one believes that many valued human activities are connected to geological and ecological features unfolding at different scales, then it makes sense to evaluate impacts on multiple scales. So, in contrast with Solow's approach to evaluating progress toward sustainability by simply calculating effective investments judged in present-value dollars, a pluralist can pay attention to multiple scales and can then choose projects and investments with both immediate impact today and neutral or positive impact on the slower-changing processes that will support opportunities indefinitely.

It may seem that I have beaten up on economists, and I admit to aggressively criticizing one role often given them—as providing a decision framework complete with quantification of costs and benefits. I have insisted that they should enter the public debate on equal terms with other stakeholders. They cannot dictate the terms of the debate or define terms like "value" as they please. While I place economic analysis in a less central place in the initial decision-making process insofar as the decisions regard broad values and social goals are concerned, I give economists an extremely important role. This role has two aspects, one creative and the other more evaluative. If one thinks of economics as the study of incentives, economists studying incentives can suggest tools and (what Heberlein calls) structural changes that will change behavior and instigate new norms. Economists can accept the role of identifying policies affecting incentives that will shift human behavior toward more environmental protection. Once social goals are set politically, economic analysis and reasoning is necessary to understand behavior (behavioral economics), to understand political institutions (institutional economics), and to understand incentive structures that can be agreed on as promising experiments in becoming more sustainable. The other, related aspect is that economics—through the activities just mentioned, as well as studies of present preference—can adopt the role of evaluator of the efficiency of proposed policies both prospectively and subsequent to their becoming on-the-ground experiments. In other words, if one frees economics from the burden of setting goals for the society, then it naturally assumes the role of cost-effectiveness analysis (CEA) with respect to means to goals set politically. In this role, economists cooperate by proposing creative incentive structures that can encourage social goals, even as they evaluate both proposed and experimental policies in search of the least costly way of achieving those goals.

In effect, this means working within a framework that involves cost analysis, but it assumes that politically supported policies will determine social and sustainability goals, thus focusing attention on how to achieve politically articulated social goals, including movement toward more sustainable living, and on how to do so more efficiently. Shifting to a more pragmatic, decision- and action-focus encourages all the disciplines to enter the discussions of what to do on equal terms. No discipline—neither environmental economics nor environmental ethics nor any other field—dictates priority goals and values. When a community is functioning effectively in a search for a sustainable path forward, representatives of all the disciplines will be embedded within the decision context, and all will contribute to public deliberation.

When the focus is on decisions and how to change behavior, rather than on utopian ideals, ethicists will engage in analyses of what moral virtues are appropriate in particular situations, providing abstract but locally placed arguments about what is right and wrong in a particular situation. Social scientists, including economists, will be engaged in seeking both structural changes and incentive changes that will encourage behavioral change and the emergence of new norms. Focusing on decisions faced in particular situations will not cause deliberators to avoid abstract argumentation about principles derived from academic disciplines; rather, these principles will take on new life as they are applied in many situations. Experimentation within an adaptive management process can breathe new life into disciplines as disciplinary ideas are discussed and applied in particular situations. What will have changed if a society embraces the adaptive approach to learning and decisions will be that, to fulfill the function I am defining for them, the academics will be active players within the deliberative and action-directed processes necessary to learn the way to sustainability.

Sustainability is not a state but a process. Perhaps it will always be tempting to speculate about what it would be like to live sustainably; and when imaginations ultimately fail to create an integrated conception, perhaps it is not surprising that those of a utopian bent fill in the enormous blanks in knowledge of the future with speculative theory. But if one takes seriously the magnitude and acceleration of human technological change, which will, in turn, interact with background changes that will be expressed at all levels and scales of the atmosphere and the biosphere in the coming decades, then the only alternative to despair is the hope that rests on developing processes that can generate better decisions. As our character Adapt has recommended, such processes must be both adaptive in correcting errors in understanding of systems through science and also capable of self-critical examination of society's ends and aspirations. Drawing on the pragmatic tradition, I have followed Dewey in thinking of values as a kind of hypotheses, open to refutation by experience, so that our examination and attempts to improve decision processes is in all respects a struggle to apply the scientific method more and more broadly in facing the future.

What is needed is a process—not one bound to narrowly defined outcomes—that keeps multiple options and opportunities open, yet is robust and flexible enough to respond to unexpected changes and to allow creative responses. Such a process cannot be rigid; it must be a process whereby its participants can learn how to learn as it faces countless new situations. The process, if it is to fulfill the promise of a truly normative conception of sustainability, must also incorporate a new, more flexible, process of evaluating change and human responses to environmental changes. Currently, most discussion of environmental values and evaluation of change takes place within academic disciplines or think tanks. In economics, in environmental ethics, and in environmental psychology, particular cases are discussed, though mainly as examples of methods or specific arguments. The application of such abstract reasoning and amalgamation of multiple academic discussions of types of values to solve actual environmental problems is left to nonspecialist participants in public processes. Or, more often, it is simply ignored. My goal here has been to show that evaluation should be embedded within actual adaptive processes of decision making. This inclusion will require professionals who possess expertise in particular methods to be embedded within the deliberative process, rather than existing in an ivory-tower world divorced from the particularities of actual environmental problems. Since most of those problems are wicked problems, there will be no algorithmic solution available based on a general theory of value. This means that professionals skilled in evaluation will be invaluable; yet their contribution will need to be integrated into a messy, ongoing process of public deliberation. Only when embedded and integrated into the discourse regarding a problem can evaluation truly address environmental problems as multilayered systems problems.

So, what is sustainability? In this last section, let me survey what we have learned about this challenging—and essential—concept that is causing so much discussion.

First, as just noted, sustainability is not a state of being but a learning process by which communities make decisions that are progressively more long-sighted as individuals and the society expand their thinking into the long present. Sustainability is not the optimization of any single variable; it is the unfolding of a new consciousness and a new ability to learn from experience and adapt to changing environments.

Of course, humans will always live in the present so, second, sustainability apparently requires a multilayered, systems understanding of the context of human life and the unfolding of human activities, an insight that can be traced to John Muir's work and to Aldo Leopold's powerful and versatile simile of learning to think like a mountain, which has "lived" so much longer than humans or human societies. I have signaled this key idea by encouraging the development of models for processes and problems using hierarchical models that encourage systems thinking. Hierarchical models provide formalisms that can create more effective ways to think about space–time relations; hierarchy theory (HT) doesn't resolve scalar problems, though it can encourage dialogue about cross-scale relationships and highlight cross-scale impacts.

Third, to evaluate whether a community is moving toward sustainability, it is important to identify and articulate features of an ecophysical system that are essential to support important human options and opportunities. Theoretically, this aspect highlights Amartya Sen's idea of "opportunity freedom," which refers to the range of choices available to an individual or to a population in the search for a meaningful life. Sen's sense of freedom undergirds the formal, schematic version of our definition, which defines sustainability as a relationship between generations that maintains opportunities.

It was noted that it is much easier to identify instances of unsustainability than of sustainability. Whenever one can foresee that certain decisions or actions threaten an opportunity valued by members of a community, a sustainability problem is faced. Characterizing some development path as positively sustainable, according to this definition, one would have to consider a wide range of opportunities, and perhaps would also have to consider whether substitutions of other opportunities might lead, overall, to a more sustainable outcome. The schematic feature of the definition is necessary because sustainability will be pursued and judged within processes taking place in particular communities. Since I do not believe that there a single definable state of sustainable living, the opportunities associated with sustainability will vary across communities as a result of that community's commitments and norms as well as its ecophysical environment. It has also been suggested that public deliberation focus on choosing indicators—easily measurable features that are taken to support valued opportunities—because coalitions may form to support indicated features that support more than one opportunity, as in synoptic indicators such as areas remaining as pervious surfaces.

Fourth, for those who might prefer a simpler, more intuitive definition, I can offer the following: sustainability of a community is represented as development and maintenance of the multigenerational public interest for that community. Sustainable policies for a community are ones that protect not the interests of individuals, however aggregated, but instead the public interest of a given community as emergent in the present and future generations. As developed in chapter 5, this version of the public interest can be defined procedurally and counterfactually: the public interest is what a community would decide if its members pursue, through enlightened, adaptive, and fair policies, the multigenerational public interest. What the community would decide under those conditions, it is hypothesized, would be to protect opportunities for the future.

Fifth, my emphasis on process, together with an even-casual observation of losses of opportunities associated with current development paths, forces a corresponding emphasis on transformation and a need for better decision making. Philosophers, economists, social scientists, and natural scientists alike can provide information, ideas, and opinions within deliberative processes. So they have an important place, embedded in debates about how to respond to those situations.

As Heberlein and other psychologists and cognitive scientists have found, however, human behavior is seldom changed by ideas and information alone. This implies that, if a community is to learn to be sustainable, then theories and scientific information will be necessary but not sufficient to lead to behavioral change. True behavioral change will occur within deliberative community procedures that are focused on actual decisions. Citizens must be involved in these processes, and they must be willing to make decisions such as those affecting incentives toward individual actions. If they institute changes in incentives, they will also change the structure of situations and affect their own and others' behavior. To this end, deliberative processes should employ the tools and methods developed in chapter 7. Deliberation must be deep enough not just to change immediate preferences but to change how problems are conceived and framed, thereby transforming the community into a learning community.

Leopold's deep insights—that impacts of human activities have effects on multiple scales and that most natural changes are slow and take place against a backdrop of relatively stable environmental conditions—has encouraged me to develop a multiscalar, systems model. Key to such models, including the ones in which humans live and act, is the recognition that larger systems normally change much more slowly than do their smaller components. This means that, when acting within a system with a short temporal frame in mind, one must inquire whether changes at the smaller scale might become so pervasive as to set in motion undesirable changes at a larger scale—the scale that supports valued opportunities.

Alternatively, scalar thinking can be helpful in identifying actions that can be argued to have positive impacts on all scales, viewed from some point in a larger system. One encouraging sign of progress toward sustainability in less-developed countries can be found in examples of tree planting programs, which have been successful in Africa and Asia.[1] These projects involve indigenous people, usually women, who are aided by a nongovernmental organization (NGO). The NGO provides women with startup resources to permit them to plant and tend trees on degraded and unused land. Once rights to use the fallow land are instated, the women are furnished with tree seedlings and given instructions to plant them close together. Within a short time, the growing trees will need thinning, which provides the women with wood for their own use and for sale. As the trees grow, being thinned each year and providing income, the trees stabilize soil and improve soil quality. Note that such an investment has benefits on multiple scales: early culling of crowded seedlings provides firewood and income for the tree growers, the trees have positive ecological effects on soil, and the growing trees capture carbon from the atmosphere, affecting global climate. As an extra bonus, research on population growth has shown that providing economic opportunities for women is one of the most potent factors in reducing family size and population growth, which reduces pressure on world resources. Investments

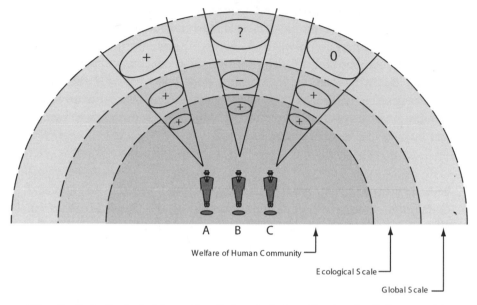

Figure E.1. Scalar Pareto decision making: Having adopted a multiscalar, pluralistic approach to value-based decision making, it is possible to summarize the appropriate decision process as encouraging actions that are positive on all scales or, if no such policy is available, to choose policies that show positive effects on at least one level and no serious negative impacts on other levels.

such as this have almost immediate economic impacts and, in the longer term, positively affect the ecological health of an area and reduce the forces driving climate change. Creative solutions such as this negate Solow's paradox. Investments such as this create no conflict between helping today's poor and striving for sustainability.

Leopold's insight, as developed in detail in this book, can perhaps be captured by modifying an idea from economic theory. Economists have noted that perhaps the easiest decisions a community of deciders can make are ones in which it is possible to improve the lot of one or more subjects without harming any others. Finding such a Pareto Improvement can provide obviously rational choices in certain cases. The problem is that very few difficult problems have solutions that will harm nobody. So, Pareto thinking, while foundational to mainstream economics in an important sense, does not carry us very far when applied at the individual level.

Perhaps, however, it is possible to extrapolate from such thinking regarding individuals and apply the Pareto reasoning to a multiscalar system of analysis that recognizes that important human values are affected by processes at multiple scales in space and time, as illustrated in figure E.1. I am suggesting a scalar Pareto model for decision making. Such a model would evaluate the impacts of proposed decisions by considering their impacts on all scales—the impact on individual well-being (economics); the impact on plants, animals, and their habitats; and the impact on global systems like the atmosphere. The systems on each of these scales support important human opportunities. So I propose a scalar Pareto decision rule that judges an action in terms of whether it helps support opportunities on economic, ecological, and global scales. If no such actions are available, then communities should choose actions that will support and improve opportunities on at least one scale, without seriously damaging processes on other scales that support human opportunities. In this book, I have tried to develop and describe community-based processes capable of applying such a decision rule on the way to achieving sustainable living.

Adapt's Ten Heuristics

Heuristic 1

The process rationality heuristic (section 3.3). Pay attention to process. The rationality of decisions depends more on finding an appropriate process than on making accurate calculations and predictions.

Heuristic 2

The epistemological heuristic (section 3.6). Trust experience more than ideology; question everything, but not all at once. Be experimental. Robust approximations are better than precise projections.

Heuristic 3

The normative sustainability heuristic (section 4.2.3). Engage the community in a search for enduring values constitutive of its identity. While economic concerns and criteria are important to any community, truly normative sustainability requires a public discourse and the emergence of a "public" that can identify key values that must be sustained if the public interest is to be sustained.

Heuristic 4

The scale and boundaries heuristic (section 4.3.3). Pay attention to scale and self-organization, both in physical dynamics and in political institution building. Getting the scale right is an essential aspect of problem formulation. Whoever is affected by decisions within boundaries of the system bounded for management should be heard.

Heuristic 5

The process evaluation heuristic (section 5.3.1). Think of evaluating change as an ongoing process; the goal of evaluation should be a comparison, and an ordinal ranking, of possible development paths into the future, an activity that continues as a community transitions from present to future. A multiscalar approach allows the evolution of development paths over multiple scales of time and space, capturing the idea of pursuing a multigenerational public interest for the community.

Heuristic 6

The boundary institutions heuristic (section 5.3.3). Construct boundary objects and institutions that encourage a community of interests. For most complex problems today, there is no "spatially appropriate" institution or jurisdiction available; let emergent public voices articulate problems and bound appropriate institutions.

Heuristic 7

The sustainable indicator heuristic (section 5.4.2). Evaluate development paths according to their likelihood of achieving the public interest, conceived as a widely acceptable balance in protecting important variables—a suite of indicators deemed important enough to measure and monitor as sustainability indicators. Arguments that a given indicator is important and should be monitored express values of process participants.

Heuristic 8

The dynamic evaluation heuristic (section 6.2). Never assume values are fixed. Adaptive management provides opportunities to question and re-form values in process through social learning. Dialogue and deliberation should be inclusive, but shaping of values should aim at articulating and pursuing the emergent public interest.

Heuristic 9

The capabilities heuristic (section 6.3). Evaluate environmental change from the viewpoint of broadening the capabilities of both present and future people. Paying attention to capabilities emphasizes freedom of choice over many opportunities to achieve well-being and highlights ways in which physical processes support the multigenerational public interest.

Heuristic 10

The fitting many metaphors heuristic (section 8.4). Avoid single, master metaphors. Metaphors close off as well as open up vistas and perspectives; therefore work with a shifting kaleidoscope of metaphors that will also track shifts in language and conceptualizations. Seek integrative metaphors (e.g., Leopold's thinking like a mountain) that are more inclusive of greater visions and encourage a balance among varied social values flexible enough to evolve as situations change.

GLOSSARY

Accounting problem. The problem on which economists and others, especially environmental ethicists, disagree regarding how to understand environmental value and what types of things have it.

Adaptive management (AM). An emergent approach to protecting resources and the environment that challenged and is replacing SRM as a dominant idea in conservation, including in some government agencies. Adaptive management favors learning by doing and urges protection of system-level characteristics of natural systems (such as resilience). Based on an analogy to Darwinian selection, adaptive management has come also to embrace public involvement, as in "adaptive collaborative management."

AID (US Agency for International Development). US agency that oversees development efforts in less-developed countries. This agency has become a strong advocate for sustainable development over the past decades.

Algorithm. An algorithm is a process or formula that will, if followed, lead to a definitive outcome. One can say that a given problem has an algorithmic solution—or that it does not.

Anthropocentrism. The view that all and only human individuals have intrinsic value and that only humans count in ethical reasoning. It is the contrary of "nonanthropocentrism."

Attitude. A belief that is tied to a value—it always has an object that is judged better than something else.

Back-casting. A procedure whereby a group first defines long-term objectives and then reasons backward to identify key steps and a step-by-step timetable for achieving the objective.

Benign problems. Contrasted with "wicked problems" (see below), a benign problem is one that has a single correct answer, even if that answer is very complex and not yet known. Problems in mathematics, no matter how complex, are generally benign.

Bequest. The sum total of resources that one generation passes to the next.

Biocentrism. A version of nonanthropocentrism that places intrinsic value on individual organisms. It is sometimes contrasted with "ecocentrism," while both contrast with anthropocentrism.

Boundary critique. A feature of CST that encourages reflection on the boundaries and membranes that define a system.

Boundary institution. An institution that is helpful in communicating and fostering action across social, political, or disciplinary boundaries.

Boundary object. An object with characteristics that are sufficiently plastic to allow them to be understood and used in more than one context. Such terms must be adaptable to multiple discourses and also robust enough to be understood across bounded discourses.

Capabilities approach. An approach to evaluating outcomes whereby value emerges in situations rather than from fixed preferences. On this view, moral well-being depends on choosers having real opportunities that support their individual values.

CBA (cost-benefit analysis). A methodology used by economists and others to identify and estimate the costs and benefits associated with an action or policy, evaluated either retrospectively or prospectively. CBA can be used either as a decision rule or as an informal aid in a broader decision-making process.

CEA (cost-effectiveness analysis). An analysis and comparison of the costs of actions to achieve an accepted goal. CEA comes into play when a goal is committed to politically and the question is how to reach it at least cost.

Community. A group that shares a sense of identity. When that identity includes devotion to a natural system, a community can be understood as a "place."

Constraint. A limit resulting from resource unavailability, which limits human choices in the struggle to survive. Contrary of "opportunity."

CST (critical systems theory). The theory that system analysis must be reflective about the broader, including social, impacts of bounding and describing a system in a particular way. (See "boundary critique.")

Curiosity-driven science. Disciplinary science pursued for the sake of knowledge.

Discounting. The practice, in economic analysis, whereby present and earlier outcomes of an action or policy are counted more heavily than later ones. This is accomplished by "discounting" the impacts occurring in each year in the future by some percentage.

Distance problem. How far into the future should any given generation care about?

DMV (deliberative monetary valuation). A method of studying values of a group by eliciting, discussing, and moving toward consensus regarding an appropriate figure for a monetary valuation of a particular good or decision affecting goods.

Double-loop learning. Learning that causes a change in perspective and calls into question widely held assumptions.

EBM (ecosystem-based management). A category of large-scale and top-down management efforts discussed by author Judith Layzer, who criticizes these efforts as insufficiently protective of the environment.

Ecocentrism. A version of nonanthropocentrism that attaches intrinsic value to ecosystems and species. Often contrasted with "biocentrism" and opposed to anthropocentrism.

Ecological economics. A field of study that developed in the 1980s and 1990s as ecologists and economists began working together to determine the interrelationships of economics and ecology.

Economism. The theory, advanced by some but not all environmental economists, that every value can be represented in economic terms. (See "WTP.") When capitalized, contrasts with "economics."

Ecosystem services. Value derived by humans from the use of products and services from ecosystems; often classified as provisioning, regulatory, and cultural ecosystem services.

EIS (environmental impact statement). A detailed account of the possible and likely effects of a project, action, or policy. The US NEPA (1969) required advocates of all policies and actions of large scope to prepare an EIS, which required the holding of public hearings.

Endangered Species Act of 1973. A landmark in environmental legislation in the US, the ESA protects species listed as threatened or endangered against further endangerment.

Epistemology. The theory and study of knowledge and how it is obtained.

Equilibrium systems. Any system that is described to fluctuate around fixed points: the system is expected to equilibrate at the prior set point after a disturbance. Such systems must be closed to the systems outside them.

Fair savings rate. The rate of consumption of a society that does not reduce the accumulated capital of a society. A generation achieves a fair savings rate if it produces what it consumes and does not reduce the general stock of capital.

Fallibilism. The view that every belief is open to question. Denies foundationalism.

Fatalism. Generally, the idea that events are determined in advance by some power; in this book it refers to an attitude that nothing can be done about a situation such as a currently misleading set of definitions.

First-generation environmental problems. Early environmental problems that were local in nature, such as pollution in a particular spot or local overhunting of a game species; contrasted with second- and third-generation problems.

Flips. An informal way of referring to major shifts in the functioning of an ecological system. Reductions in the resilience of ecological systems can lead to flips, which can be economically disastrous for communities dependent on the systems.

Foundationalism. The view that true and certain knowledge can only be achieved if there are some propositions that are known noninferentially.

Futurism. A position that favors the interests of future persons more than the present persons, advocating sacrifices by present people for the well-being of future people.

FWS (Fish and Wildlife Service). The agency in the US Department of Interior devoted to protecting and managing wildlife. This agency administers the US Endangered Species Act of 1973.

GREM (grassroots ecosystem management). A category, introduced by Judith Layzer, of locally initiated (bottom-up) processes to protect natural systems and features.

GST (general systems theory). The interdisciplinary study of the general structure of systems, including their ability to self-regulate and self-organize through feedbacks.

Hammer. A threat by a governing body to create a bad situation for an area or community if it fails to address a problem deemed important by the governing body.

Heuristics. Generally, rules of thumb that help when making decisions. Here, guidelines to help communities improve communication and make better decisions.

Heuristic proceduralism. Respects *appropriate* procedures as the path to sustainability; this version of proceduralism treats decision processes as rational if they follow procedures recommended by a set of important heuristics.

HT (hierarchy theory). A theory in ecology—often associated with GST—that defines a collection of models possessing two characteristics: all observations and measurements are taken within the system, and smaller subsystems are expected to change more rapidly than encompassing systems that form their environment. HT does not assert a single organization as the correct one for a system; rather, it provides a useful formalism that provides choices in creating models of ecological phenomena.

Ideology. Generally used to refer to broad ideas and worldviews. In this book, it refers more specifically to broad ideas that cannot be falsified by observation or experience.

Ignorance. A hypothesis that people in any given present time cannot know what resources people of the future will need. This implies that sustainability cannot be understood as providing future people what they will want or need.

Instrumental value. The value that a nonhuman object provides to someone or some thing, as in the value of ripe fruit to a hungry person.

IPP (in principle patch). A ploy often used by advocates of algorithmic solutions and of one right answer to complex problems, when they assume that currently unavailable information will eventually "fill in the blanks" and lead to a single, desired outcome.

IVT (intrinsic value theory). The theory, favored by many environmental ethicists, that some elements of nature have intrinsic value—a good of their own that must be considered in ethical decision making. Often contrasted with "instrumental value."

Jeopardy. When US FWS finds that an action or policy threatens an already endangered species, it makes a declaration of jeopardy and requires advocates of those actions to find a "reasonable and prudent alternative" to current or proposed activities.

Limited realism. The epistemological view that human experience originates outside the person (realism), but that the objects of knowledge are relative to linguistic conventions.

Macroscoping. A mental process of shifting attention from a limited system to a larger one, as when a river is reconceived as an element of a watershed.

Marginal value. The highest value a consumer will pay for an object, which means that aggregations of value based on consumer WTP represent the value of the object "at the margin."

Mission-oriented science. Science that is pursued within public discourse and in controversial social situations.

Monism. A category of theories characterized by commitment to a single principle that is believed by its advocates to account for all aspects of a particular area of knowledge. In this book, it refers to ethical monists, who believe all ethical quandaries can be resolved by a single theory or rule, and by Economists, who believe all value questions are amenable to analysis by economic principles (opposed to "pluralism").

Natural capital. Refers to features of a natural system that should, according to its advocates, be protected by each generation in order not to harm subsequent ones.

NEPA (National Environmental Policy Act of 1969). The law that established a US national policy of protecting and enhancing environmental quality; it included procedural requirements that required public input into decisions affecting environmental systems and human health.

NGO (nongovernmental organization). Any organization that works independently of governments to address problems within its purview.

Nonanthropocentrism. The view that at least some elements of nature have intrinsic value. Contrary to "anthropocentrism."

Noncomputability. When a problem has no algorithmic solution, it can be said to be "noncomputable."

Nonpoint-source pollution. Pollution that occurs across whole areas such as agricultural run-off or small amounts of pollution from countless sources. Contrasted with point source pollution.

Norm. A behavioral regularity established by habit.

Normative sustainability. A strong form of sustainability that emphasizes the role of values and social learning about them in conjunction with concern for ecological systems.

Open systems. Systems that are separated from their environment by permeable membranes. An open system must be understood in the context of the system that surrounds it.

Opportunity. An open human choice based on an available resource base. "Opportunity," here, is a bridge concept that has empirical content about availability of resources necessary to allow individuals to choose actions valued by them. Contrary of "constraint."

Pareto Improvement. An action that will improve the lot of at least one person and harm nobody else. The Pareto Improvement criterion urges actions that produce a Pareto Improvement; unfortunately, this criterion applies to few actual cases because, in complex political situations, there are no actions that will harm no one.

Performative. An act of ethical commitment. In general, an utterance that establishes a responsibility, as in marriage vows. Here, the act of a community to commit itself to protect (sustain for the future) its deepest values and aspirations.

Pervious surfaces. Surfaces and soils that allow ready penetration of liquids.

PES (payment for ecosystem services). Assignments of an estimated monetary value to human use of ecosystem services.

Pluralism. The view that complex phenomena can be understood in multiple ways, depending on one's purposes or perspective.

Point-source pollution. Pollution coming from an identifiable source such as a sewage outflow or a particular factory.

PP (precautionary principle). A decision rule that advises caution in situations of high uncertainty and high stakes. It is often said to shift the burden of proof onto developers and others who propose large and risky projects and policies.

Pragmatism. A philosophy that developed in the United States in the latter decades of the nineteenth century. It draws heavily on Darwinism and emphasizes community experience, science, and action. It is argued here that it provides a philosophical foundation for "adaptive management."

Preferences. Individual preferences are the predilections that explain individual choices. Economists measure preference, which they take to be unchanging and existing prior to decisions, as willingness-to-pay (WTP).

Presentism. The view that people of the present should not, or need not, act to protect the well-being of future people.

Procedural rationality. A decision is rational if it was arrived at by an appropriate procedure.

Public. Used in the sense introduced by John Dewey: a public emerges as citizens become aware of threats to their well-being and as members coalesce and create a force in the public sphere.

Public interest. The good of a whole community, contrasted with the special interests of individuals, even when these are aggregated across the whole community.

Rational choice models. Decision models that set out to identify what is good for a population by aggregating individual goods—understood as maximization of individual advantage—to arrive at a social good.

RDS (risk decision square). Defines a decision space ordered by plotting the degree of reversibility (a temporal variable) against a variable of spatial extent of impacts to determine the appropriate decision rule for a given problem.

Reductionism. An overemphasis on analysis whereby systems and objects are understood in terms of their smallest parts, thereby denying that encompassing systems have emergent qualities that cannot be explained by reference to constituent parts.

Resilience. The characteristic of systems that have the ability to return to normal functioning after being disturbed.

Scenario development. A group exercise in examining possible futures and how society might achieve a desirable outcome.

Schematic definition. A definition that provides the general structure of an adequate definition but leaves open decisions regarding how that definition will be completed through a community's deliberation about its deepest values and dearest aspirations.

Second-generation environmental problems. Problems that cross regional or national boundaries, such as acid rain or extirpation of a species over its entire range. Contrasted with "first-" and "third-generation" environmental problems.

SMS (safe minimum standard of conservation). A decision rule that advises to save the resource, provided the social costs are bearable. SMS is favored in situations of high uncertainty and high risk of irreversibility (as in threats of extinction of a species). It is often associated with CEA, which counts and compares costs while assuming the resource has positive value and is an alternative to CBA.

SRM (scientific resource management). A widely employed approach to resource use that developed as one approach to "conservation" in the United States in the nineteenth and early twentieth centuries. SRM emerged as the favored approach to conservation in government agencies. Often associated with the ideas of Gifford Pinchot, the first chief of the US Forest Service, this idea remained dominant in the agencies until the 1960s and 1970s. Here, it is contrasted with "adaptive management."

Strong sustainability. A version of sustainability that, in addition to maintaining the accumulated capital of the society, requires each generation that it protect certain aspects of nature, sometime referred to as "natural capital."

Stronger economic sustainability. A version of sustainability that supplements an obligation to maintain the society's stock of economic capital by also protecting "natural capital."

Stuff. A term defined to encompass descriptions of nonwelfare goods that are designated for protection in versions of "strong sustainability."

Substantive rationality. A decision is substantively rational if it contributes to the attainment of stated objectives.

Substitutability. The extent to which alternative resources can be substituted for other resources that are becoming scarce; in particular, can human-made resources substitute broadly for natural resources?

Third-generation environmental problems. Problems with a global scope, such as climate change and the worldwide biodiversity crisis.

Tradeoffs. Sustainability seems to involve many choices between current consumption and investing for the future. Should the present generation sacrifice to improve the lives of future people who may be richer than the present?

Transformative. A linguistic trope, such as a metaphor, analogy, or simile, that has the potential to change one's perspective.

Typology of effects. How can one tell whether a given action involves an issue of sustainability? For example, how might one determine whether a given pattern of tree foresting is "sustainable"?

Utilitarianism. The ethical theory that "good" should be understood in terms of the aggregated happiness of a society or group. There are multiple versions of this theory, as happiness is defined differently; economics employs a specific version of utilitarianism that equates happiness with preference fulfillment.

Weak sustainability. A version of sustainability that equates it with the maintenance across generations of the accumulated wealth of a society, which is achieved by maintaining a "fair savings rate."

Wicked problems. Problems that have no definitive solution, often because they are not formulated in a way accepted by partisans of different positions. Most environmental problems are wicked problems or have wicked elements.

WTP (willingness-to-pay). Economists' measure of the value of a thing as the price a consumer is willing to pay for it. This price is the amount a consumer would pay for the last such item of a type that they purchase, which means that economic values are calculated "at the margin."

NOTES

PREFACE
1. I will use the term "community" in two senses, recognizing the ambiguity of this term but at the same unable to supply a good alternative. First, as here, I will use "community" very loosely to mean any group of people who share a sense of identity. In this very broad sense, the term could refer to a geographically noncontiguous group that forms on the Internet or to a temporary group who comes together for a specific purpose. When "community" in this loose sense refers to a group of people who live in a particular place and who identify with their ecophysical surroundings, we refer to such a community as a "place" (composed of people and their relationship to a specific geographic area). Later, we will use the term "community" more precisely to refer to a "public" that has, through effective institutions, united behind concerns that bind them as a political community.
2. Fortunately, there is a very good book that tackles addresses these problems (Ostrom, 2007).

CHAPTER 1
1. It is true that European environmentalists have made some valiant efforts in this direction, yet the conflicts between different conceptions of environmental value have not been resolved even in these efforts, leaving the discourse of evaluation fractured.
2. Mark Sagoff has applied Archilochus's distinction to environmentalists, though he uses it slightly differently than I do. For him, the hedgehog sees all environmental problems as having a single cause, while the fox takes problems as they come (Sagoff, 2002).
3. It should be said, in fairness to the commission, that the text, if not its definition, does discuss concerns about ecological limitations on the use of resources.

CHAPTER 3
1. I use the capitalized forms "Economism" and "Economist" to distinguish from the lower-case versions. When capitalized, the term refers not to all economic theories or all economists but solely to a subset of economists—those who endorse the view that all environmental values can, in principle, be expressed in economic terms.
2. Other approaches, such as risk analysis, contractualism, and the like, represent examples of alternative attempts to create comprehensive modeling of complex problems.
3. See National Research Council, 1996, and figure 6.1.

4. These individuals will appear in the story below, complete with detailed references to their work.
5. Kai Lee was the first to recognize that John Dewey offered many ideas that can support adaptive management (Lee, 1993).
6. Emphasis on local places can raise the much-discussed concern about NIMBYism (not in my backyard). There is no doubt that, in many cases, local people, while recognizing that an enterprise such as a landfill is necessary, reflexively mount opposition to it. While I recognize that this reaction can be irrational, I also think that this instinct to protect one's home place represents a healthy concern for one's home. (See Norton and Hannon, 1998.)

CHAPTER 4

1. Under this meaning, I would *not* exclude, for example, a far-flung "Internet community" under my definition. Such a community, however, has no spatial connection to a geographical area. For a community to become a true place, its identity must be connected to a geographic location. The definition might not exclude a worldwide "community" and might identify with and protect the "place" that is the earth and its atmosphere.

CHAPTER 5

1. Ostrom (1990, 2007) has studied and discussed in detail community-based management schemes and the patterns that affect their success. I will return to this point in chapter 9.
2. I learned of the interest and importance of building synoptic indicators in work I did with Anne Steinemann. I am grateful to her for sharing this idea (Norton and Steinemann, 2001).
3. Of course, one must also care about substantive impacts of choices on the environment; see section 9.2 for a discussion of the difficulties in directly studying these impacts.

CHAPTER 6

1. One upshot of this tendency toward "chunking" is the associated practice of creating inventories of various types, such as creating inventories of species, of endangered species, of endangered habitats, and so on. While these inventories may be useful in some contexts, it is always important to remember that all these entities exist within changing processes that can be more important than any particular listed entities.
2. My reasoning here is similar to that of Breena Holland (2012), who proposes supplementing the capabilities approach by introducing the idea of "environmental meta-capabilities" to capture the idea of the dependence of capabilities on ecophysical characteristics of the environment in which individuals act. However, I believe this essential idea can be included in the third category of environmental conversion factors by connecting those conversion factors to my idea that ecological systems present themselves to individuals as opportunities, with "opportunities" defined to include physical processes and features that support those opportunities.

CHAPTER 8

1. See Norton (2005) for a discussion of economists' proposals to avoid tragedies by privatizing common resources.
2. See also Taylor, 2005; Larson, 2011.

CHAPTER 9

1. For more exhaustive reviews, see Leach et al., 2002; Sabatier et al., 2005; Wondolleck and Yaffee, 2000.

2. Apparently, some groups use more than one decision rule, as the percentages offered add up to more than 100.

3. The account of the Platte River negotiations and agreement follows that of David M. Freeman in his outstanding book, *Implementing the Endangered Species Act on the Platte Basin Water Commons* (Freeman, 2010).

4. I refer to a National Science Foundation grant under which we asked the question, how do individuals and groups bound the systems to which they attribute environmental problems? "Ecological Boundary-Setting in Mental and Geophysical Models," National Science Foundation, Grant #0433165, funded by SBE-HSD, 2004–2008.

CHAPTER 10

1. I cannot fault Callicott for failing to respond to the paper I wrote with Hirsch, even though it anticipated his title, "Learning to Think like a Planet," and also predated publication of his book, which may well have been in production by the time our paper was published. I respond to Callicott's book here in order to clarify our agreements and disagreements. In this note, I will briefly respond to a few of the criticisms of my work on Leopold as highlighted in his new book; these responses can be brief, since Callicott repeats arguments made before, and since I have recently responded in great detail to all his criticisms of my interpretation of Leopold (Norton, 2013).

   Many of the issues between Callicott and me have to do with whether Leopold, who unquestionably underwent a major transformation of thought beginning around 1923, shifted from a monistic, economic understanding of humans and environment to the other extreme, a radical and uncompromising anti-anthropocentrism, or whether (as I argue) he broadened and deepened his understanding by adopting a pluralistic approach that integrates human concerns with concerns for "the mountain." Our disagreements, as will be obvious to readers of Callicott's book, focus specifically on how to interpret Leopold's discussion of conservation ethics in the last section of the essay titled "Some Fundamentals of Conservation in the Southwest," which was posthumously published in 1979. As I read this essay, Leopold (who in the earlier sections of that essay was highly critical of land management in the Southwest) was surveying the various approaches to providing an ethical critique of these unfortunate actions by his agency, which he observed to be disastrous both economically and ecologically. On my reading, Leopold explores several approaches, starting with a biblical reference to the prophet Ezekiel, who criticized bad agricultural practices; moved on to discuss ideas of the Russian thinker P. D. Ouspensky, who advocated a robust form of organicism; and mentioned John Muir, to whom he attributed the view that rattlesnakes have rights.

   Leopold clearly saw the difficulties and ironies involved in an unquestioned anthropocentrism such as is embodied in most economic thought. At a crucial juncture in his argument, Leopold notes that both religion and science (referring mainly to applied sciences such as forestry) adopt an anthropocentric stance. In a crucial passage, regarding which Callicott and I strongly differ, Leopold refers to this assumed anthropocentrism and writes, "Since most of mankind today profess either one of the [anthropocentric] religions or the scientific school of thought which is likewise [anthropocentric], I will not dispute the point." I take Leopold to be saying that, even on the assumption of anthropocentrism, he can criticize bad land use, so the difference between anthropocentrism and its denial are not crucial to his case. This interpretation, which follows naturally from his use of the word "since," implies that, given its wide acceptance, it makes sense to proceed with his criticism of bad land use on the less controversial (though problematic) foundation of anthropocentrism. Callicott's interpretation, ignoring the clear implication of "since" in this sentence, understands the sentence as saying he won't bother to reason with or criticize the anthropocentrists. If that had been his meaning, Leopold would have said something like, "Since religious and scientific believers in anthropocentrism are clearly wrong, I will no longer consider such a position."

What he does, instead, is to note the clear ironies involved in an unquestioning anthropocentrism, but then he goes forward to discuss the anthropocentric position on its own terms. Callicott's claim that my interpretation fails because it does not recognize irony in Leopold's writing is simply mistaken. In fact, I correctly interpret him as recognizing the ironies of anthropocentrism and, especially, the irony that multiple cultures had lived in the Southwest—all claiming the earth was made for them—but they are all gone now and only the earth remains. Leopold is making a profound point: even if members of a culture think the world was made for them and that they are the most "noble" species on earth, the evaluation of their land-use activities can be measured in terms of long-term survival, as would be expected if one accepts Darwinian thought. Leopold makes this point by quoting Arthur Twining Hadley, who was president of Yale University when he was a student there: "Truth is that which prevails in the long run" (Leopold, 1979, p. 141). His point was that what people believe is not important: what are important are the practices in which the culture engages. Prior cultures "left the earth alive, undamaged," and Leopold immediately used their experience of surviving for many generations as the basis for an anthropocentric critique of current practices he had criticized in the earlier parts of the essay, writing: "Is it possibly a proper question for us to consider what [future cultures] will say about us?" (Leopold, 1979, p. 141).

Then, Leopold brilliantly offers an ad hominem argument applicable to all the anthropocentrists around him: "If we are logically [anthropocentric], yes" (i.e., if we are consistent in our claim that humans are the pinnacle of creation and the noblest of creatures, then our current activities can be criticized because they are not sustainable). In the last sentences of the essay, Leopold (applying Hadley's criterion of truth and right action, that it will be marked by long-term survival) suggests that our current activities are not sustainable—which creates a reductio ad absurdum argument against any "noble" anthropocentrist who overgrazes and destroys pasture and who degrades watercourses. If anthropocentrists claim nobility for humans, "by what token shall it be manifest? By a society decently respectful of its own and all other life, capable of inhabiting the earth without defiling it? Or by a society like that of John Burrough's potato bug, which exterminated the potato and thereby exterminated itself" (Leopold, 1979, p. 141).

In my interpretation, then, Leopold moved, beginning in 1923, from an unquestioning commitment to managing natural systems for the good of humans, to a more thoughtful, pluralistic position on which he could claim that current actions and policies in the Southwest could be criticized as damaging to nature, which can be viewed as a living thing, and on the grounds of the hypothesis of human nobility. On this view, Leopold's 1923 essay can be read as a survey of possible conservation ethics, with commentary, of which the conclusion is that the degrading activities of Western culture on the ecosystems of the Southwest are wrong on multiple ethical grounds.

This interpretation explains how Leopold, through the rest of his life, often shifted from anthropocentric to nonanthropocentric arguments for conservation, depending on context and on whom he was addressing. This is much more consistent with our knowledge of Leopold as an engaged progressive who worked with other conservationists and progressives, engaging in many open-ended discussions of what is good for nature and what is good for humans. It contrasts with Callicott's view of him as a "radical" who was dismissive of other scientists and conservationists, unwilling even to consider views different from his own. Adopting my interpretation would also be helpful to Callicott, relieving him of the embarrassment of having, on the one hand, to ridicule and dismiss anthropocentric elements in the land ethic, only to be forced to embrace this "ridiculous" position in order to create an ethical foundation for his earth ethic.

EPILOGUE

1. See http://en.wikipedia.org/wiki/Green_Belt_Movement for more information and links.

REFERENCES

Ackoff, R. L. 1979. "The Future of Operational Research Is Past." *Journal of the Operational Research Society* 30:93–104.

Aldred, J. 2001. "Citizens' Juries: Discussion, Deliberation, and Rationality." *Risk Decision and Policy* 6:71–90.

Allen, T. F. H., and T. B. Starr. 1982. *Hierarchy: Perspectives for Ecological Diversity*. Chicago: University of Chicago Press.

Arias-Maldonado, M. 2012. *Real Green: Sustainability after the End of Nature*. Surrey, UK: Ashgate.

Austin, J. L. 1962. *How to Do Things with Words*. New York: Oxford University Press.

Beckerman, W. 1994. "Sustainable Development: Is It a Useful Concept?" *Environmental Values* 3:191–209.

Beckerman, W. 1995. "How Would You like Your 'Sustainability,' Sir? Weak or Strong? A Reply to My Critics." *Environmental Values* 2:167–79.

Bernstein, Richard J. 1997. "Pragmatism, Pluralism, and the Healing of Wounds." In L. Menand (ed.), *Pragmatism: A Reader*. New York: Vintage.

Berrens, Robert P. 2001. "The Safe Minimum Standard of Conversation and Endangered Species: A Review." *Environmental Conservation* 28:104–16.

Bishop, R. C. 1978. "Endangered Species and Uncertainty: The Economics of a Safe Minimum Standard." *American Journal of Agricultural Economics* 60:10–18.

Blatter, J., and H. Ingram. 2001. *Reflections on Water: New Approaches to Transboundary Conflicts and Cooperation*. Cambridge, MA: MIT Press.

Blomquist, W., and H. M. Ingram. 2003. "Boundaries Seen and Unseen: Resolving Transboundary Groundwater Problems." *Water International* 28:162–69.

Bockstael, N. E., A. M. Freeman III, R. Kopp, P. Portney, and V. K. Smith. 2000. "On Measuring Economic Values for Nature." *Environmental Science and Technology* 34:1384–89.

Bozeman, B. 2007. *Public Values and the Public Interest: Counterbalancing Economic Individualism*. Washington, DC: Georgetown University Press.

Bozeman, Barry, and Ben Minteer. 2007. "Toward a Pragmatic Public Interest Theory." In B. Bozeman (ed.), *Public Values and the Public Interest: Counterbalancing Economic Individualism*. Washington, DC: Georgetown University Press.

Brand, S. 2010. *Whole Earth Discipline: Why Dense Cities, Nuclear Power, Transgenic Crops, Restored Wildlands, and Geoengineering Are Necessary*. New York: Viking.

Burke, E. 1910. *Reflections on the French Revolution and Other Essays*. Covent Garden, UK: JM Dent & Sons.

Callicott, J. B. 1989. *In Defense of the Land Ethic*. Albany, NY: State University of New York Press.

———. 1992. "Rolston on Intrinsic Value: A Deconstruction." *Environmental Ethics* 14:129–43.

———. 1995. "Environmental Ethics *IS* Environmental Activism: The Most Radical and Effective Kind." In D. E. Marietta and L. Embree (eds.), *Environmental Philosophy and Environmental Activism*. Lanham, MD: Rowman and Littlefield.

———. 2014. *Thinking like a Planet: The Land Ethic and the Earth Ethic*. New York: Oxford University Press.

Ciriacy-Wantrup, S. V. 1963. *Resource Conservation: Economics and Policies*. Berkeley: University of California, Division of Agricultural Sciences, Agricultural Experiment Station.

The Citizen's Handbook. 2013. http://www.citizenshandbook.org.

Clark, C. W. 1973. "The Economics of Overexploitation." *Science* 181:630–34.

Clark, W. C., and N. M. Dickson. 2003. "Sustainability Science: The Emerging Research Program." *Proceedings of the National Academy of Sciences* 100:8059–61.

Cochran, M. 1999. *Normative Theory in International Relations: A Pragmatic Approach*. Cambridge, UK: Cambridge University Press.

Costanza, R. 1991. *Ecological Economics: The Science and Management of Sustainability*. New York: Columbia University Press.

Costanza, R., R. D'Arge, R. De Groot, S. Farber, M. Grasso, B. Hannon, and K. Limburg. 1997. "The Value of the World's Ecosystem Services and Natural Capital." *Nature* 387:253–60.

Daily, G. C., ed. 1997. *Ecosystem Services: Benefits Supplied to Human Societies by Natural Ecosystems*, vol. 2. Washington, DC: Ecological Society of America.

Daly, H. 1995. "On Wilfred Beckerman's Critique of Sustainable Development." *Environmental Values* 4:49–55.

Daly, H. E., and J. B. Cobb. 1989. *For the Common Good: Redirecting the Economy toward Community, the Environment and a Sustainable Future*. Boston: Beacon Press.

Darwin, Charles. 1859–1964 (facsimile edition). *The Origin of Species by Means of Natural Selection*. Cambridge, MA: Harvard University Press.

Dewey, John. 1927/1982. *The Public and Its Problems*. In J. A. Boydston and B. A. Walsh (eds.), *John Dewey: The Later Works*, vol. 2: 1925–1927. Carbondale, IL: Southern Illinois University Press.

Einstein, Albert. n.d. "Quotation #2927." Accessed February 26, 2012. http://www.quotationspage.com/quote/2927.html.

Eisenberg, C. 2011. *The Wolf's Tooth: Keystone Predators, Trophic Cascades, and Biodiversity*. Washington, DC: Island Press.

Environmental Protection Agency. 2011. *Chesapeake: Introduction to the Issues and Ecosystems*. http://www.epa.gov/oaqps001/gr8water/xbrochure/chesapea.html.

Finch, E. 2006. "Great Dane and Chihuahua." http://en.wikipedia.org/wiki/Image:IMG013biglittledogFX_wb.jpg.

Fishkin, J. S. 1991. *Democracy and Deliberation: New Directions for Democratic Reform*. New Haven: Yale University Press.

Flader, S. L., and J. B. Callicott. 1991. *The River of the Mother of God and Other Essays by Aldo Leopold*. Madison: University of Wisconsin Press.

Freeman, D. M. 2010. *Implementing the Endangered Species Act on the Platte Basin Water Commons*. Boulder: University Press of Colorado.

Funtowicz, S. O., and J. R. Ravetz. 1990. *Uncertainty and Quality in Science for Policy*. New York: Springer.

———. 1993. "Science for the Post-normal Age." *Futures* 25:739–55.

Gardiner, S. M. 2011. *A Perfect Moral Storm: The Ethical Tragedy of Climate Change*. New York: Oxford University Press.

General Accountability Office. 2005. *Chesapeake Bay Program: Improved Strategies Are Needed to Better Assess, Report, and Manage Restoration*. Washington, DC: US Government. GAO-06-96.

Giere, R. N., and B. Moffatt. 2003. "Distributed Cognition: Where the Cognitive and the Social Merge." *Social Studies of Science* 33:301–10.

Global Footprint Network. 2012. "Footprint Basics." Accessed May 20, 2013. http://www.foot printnetwork.org/en/index.php/GFN/page/footprint_basics_overview.

Gomez-Baggethun, E., R. De Groot, P. L. Lomas, and C. Montes. 2010. "The History of Eco-system Services in Economic Theory and Practice: From Early Notions to Markets and Payment Schemes." *Ecological Economics* 69:1209–18.

Gunderson, L. H., and C. S. Holling, 2001. *Panarchy: Understanding Transformation in Human and Natural Systems*. Washington, DC: Island Press.

Guston, D. H. 2001. "Boundary Organizations in Environmental Policy and Science: An Introduction." *Science, Technology, and Human Values* 26:399–408.

Habermas, Jurgen 1984. *The Theory of Communicative Action*. Trans. T. McCarthy. Cambridge, UK: Polity.

Hardin, G. 1968. "The Tragedy of the Commons." *Science* 162:1243–48.

Hajer, M. 2003. "Policy without Polity? Policy Analysis and the Institutional Void." *Policy Sciences* 36:175–95.

Heberlein, T. A. 2012. *Navigating Environmental Attitudes*. New York: Oxford University Press.

Hirsch, P., and B. G. Norton. 2012. "Thinking like a Planet." In A. Thompson and J. Bendyk-Keymer (eds.), *Ethical Adaptation to Climate Change: Human Virtues of the Future*. Cambridge, MA: MIT Press.

Holland, A. K. 1994. *The Ethics of Conservation Report Presented to the Countryside Council for Wales*. Lancaster University, UK: Department of Philosophy.

Holland, B. 2012. "Environment as Meta-capability: Why a Dignified Human Life Requires a Stable Climate System." In A. Thompson and J. Bendyk-Keymer (eds.), *Ethical Adaptation to Climate Change: Human Virtues of the Future*. Cambridge, MA: MIT Press.

Holling, C. S. 1973. "Resilience and Stability of Ecological Systems." *Annual Review of Ecology and Systematics* 4:1–23.

———. 1992. "Cross-scale Morphology, Geometry, and Dynamics of Ecosystems." *Ecological Monographs* 62:447–502.

———. 1996. "Engineering Resilience versus Ecological Resilience." In P. C. Schulze (ed.), *Engineering within Ecological Constraints*. Washington, DC: National Academy Press.

Holmes, O. W. 1997. "Natural Law." In L. Menand (ed.), *Pragmatism: A Reader*. New York: Vintage.

Horton, T. 1987. "Remapping the Chesapeake." *New American Land* 7:7–26.

Howarth, R. B., and M. A. Wilson. 2006. "A Theoretical Approach to Deliberative Valuation: Aggregation by Mutual Consent." *Land Economics* 82:1–16.

Hutchins, E., and T. Klausen. 1996. "Distributed Cognition in an Airline Cockpit." *Cognition and Communication at Work* 15–34.

Innes, J. E., and D. E. Booher. 2000. "Indicators for Sustainable Communities: A Strategy Building on Complexity Theory and Distributed Intelligence." *Planning Theory and Practice* 1:173–86.

———. 2010. *Planning with Complexity: An Introduction to Collaborative Rationality for Public Policy*. Oxon, UK: Routledge.

Jacobs, Michael. 1999. "Sustainable Development as a Contested Concept." In A. Dobson (ed.), *Fairness and Futurity*. Oxford, UK: Oxford University Press.

Johnson, M. 1993. *Moral Imagination: Implications of Cognitive Science for Ethics*. Chicago: University of Chicago Press.

Kelman, S. 1981. "Cost-Benefit Analysis: An Ethical Critique." *Regulation* 5:33–41.

Kempton, W., and J. Falk. 2000. "Cultural Models of Pfiesteria: Toward Cultivating More Appropriate Risk Perceptions." *Coastal Management* 28:273–85.

Kempton, W., J. S. Boster, and J. A. Hartley. 1999. *Environmental Values in American Culture*. Cambridge, MA: MIT Press.

Keough, H. L., and D. J. Blahna. 2006. "Achieving Integrative, Collaborative Ecosystem Management." *Conservation Biology* 20:1373–82.

Kolasa, J., and S. T. A. Pickett. 2005. "Changing Academic Perspectives of Ecology: A View from Within." In E. A. Johnson and M. J. Mappin (eds.), *Environmental Education and Advocacy: Changing Perspectives of Ecology and Education*. Cambridge, UK: Cambridge University Press.

Larson, B. 2011. *Metaphors for Environmental Sustainability: Redefining our Relationship with Nature*. New Haven: Yale University Press.

Layzer, J. A. 2008. *Natural Experiments: Ecosystem-based Management and the Environment*. Cambridge, MA: MIT Press.

Leach, W. D., N. W. Pelkey, and P. A. Sabatier. 2002. "Stakeholder Partnerships as Collaborative Policymaking: Evaluation Criteria Applied to Watershed Management in California and Washington." *Journal of Policy Analysis and Management* 21:645–70.

Lee, K. N. 1994. *Compass and Gyroscope: Integrating Science and Politics for the Environment*. Washington, DC: Island Press.

Leopold, A. 1939. "A Biotic View of Land." *Journal of Forestry* 37:727–30.

———. 1949. *A Sand County Almanac and Sketches Here and There*. Oxford, UK: Oxford University Press.

———. 1979. "Some Fundamentals of Conservation in the Southwest." *Environmental Ethics* 1:131–41.

Lovelock, J. E. 1989. *The Ages of Gaia*. Oxford, UK: Oxford University Press.

McShane, T. O., P. D. Hirsch, T. C. Trung, A. N. Songorwa, A. Kinzig, B. Monteferri, and S. O'Connor. 2011. "Hard Choices: Making Tradeoffs between Biodiversity Conservation and Human Well-Being." *Biological Conservation* 144:966–72.

Meadows, D. H., G. Meadows, J. Randers, and W. W. Behrens III. 1972. *The Limits to Growth*. New York: Universe Books.

Meadows, D. 2008. *Thinking in Systems: A Primer*. White River, VT: Chelsea Green.

Meine, C. D. 1991. *Aldo Leopold: His Life and Work*. Madison: University of Wisconsin Press.

Minteer, B., and R. Manning. 1999. "Pragmatism in Environmental Ethics: Democracy, Pluralism, and the Management of Nature." *Environmental Ethics* 21:191–208.

Mitchell, S. 2009. *Unsimple Truths*. Chicago: University of Chicago Press.

Mitchell, R. C. 1989. "From Conservation to Environmental Movement: The Development of the Modern Environmental Lobbies." In M. J. Lacey (ed.), *Government and Environmental Politics: Essays on Historical Developments since World War Two*. Washington, DC: Wilson Center Press.

Muir, J. 1911. *My First Summer in the Sierra*. Boston: Houghton Mifflin.

National Research Council. 1996. *Understanding Risk: Informing Decisions in a Democratic Society*. Washington, DC: National Research Council.

Nersessian, N. J. 2006. "Model-Based Reasoning in Distributed Cognitive Systems." *Philosophy of Science* 73:699–709.

Neurath, O. 1932/1983. *Protocol Statements*. Dordrecht, Netherlands: Reidel.

Newton, J. L. 2006. *Aldo Leopold's Odyssey: Rediscovering the Author of a Sand County Almanac*. Washington, DC: Island Press, Shearwater Books.

Norton, B. G. 1987. *Why Preserve Natural Variety?* Princeton: Princeton University Press.

———. 1990. "Context and Hierarchy in Aldo Leopold's Theory of Environmental Management." *Ecological Economics* 2:119–27.

———. 1994. "Economists' Preferences and the Preferences of Economists." *Environmental Values* 3:311–32.

———. 2005. *Sustainability: A Philosophy of Adaptive Ecosystem Management*. Chicago: University of Chicago Press.

———. 2013. "Leopold, Hadley, and Darwin: Darwinian Epistemology, Truth and Right." *Contemporary Pragmatism* 10:1–28.

Norton, B. G., R. Costanza, and R. C. Bishop. 1998. "The Evolution of Preferences: Why Sovereign Preferences May Not Lead to Sustainable Policies and What to Do about It." *Ecological Economics* 24(2):193–211.

Norton, B. G., and B. Hannon. 1998. "Democracy and Sense of Place Values in Environmental Policy." *Philosophy and Geography: Philosophies of Place* 3:119–46.

Norton, B. G., and A. C. Steinemann. 2001. "Environmental Values and Adaptive Management." *Environmental Values* 10:473–506.

Norton, B. G., and M. A. Toman. 1997. "Sustainability: Ecological and Economic Perspectives." *Land Economics* 73:553–68.

Norton, B. G., and R. E. Ulanowicz. 1992. "Scale and Biodiversity Policy: A Hierarchical Approach." *Ambio* 21:244–49.

Nussbaum, M. 2006. *Frontiers of Justice: Disability, Nationality, Species Membership*. Cambridge, MA: Harvard University Press.

O'Neill, J., A. Holland, and A. Light. 2008. *Environmental Values*. New York: Routledge.

O'Neill, R. V., D. L. Deangelis, J. B. Waide, and T. F. H. Allen. 1986. *A Hierarchical Concept of Ecosystems*. Princeton: Princeton University Press.

Ostrom, E. 1990. *Governing the Commons: The Evolution of Institutions for Collective Action*. Cambridge, UK: Cambridge University Press.

———. 2007. *Understanding Institutional Diversity*. Princeton: Princeton University Press.

———. 2010. "Polycentric Systems for Coping with Collective Action and Global Environmental Change." *Global Environmental Change* 20:550–57.

Paolisso, M. 2002. "Blue Crabs and Controversy on the Chesapeake Bay: A Cultural Model for Understanding Waterman's Reasoning about Blue Crab Management." *Human Organization* 61:226–39.

Peirce, C. S. 1986. "The Doctrine of Chances." In *Writings of Charles S. Peirce: A Chronology, vol. 3 (1872–78)*. Bloomington: Indiana University Press.

Pickett, S. T. A., M. L. Cadenasso, and J. M. Grove. 2004. "Resilient Cities: Meaning, Models, and Metaphor for Integrating the Ecological, Socio-economic, and Planning Realms." *Landscape and Urban Planning* 69:369–84.

Pickett, S. T., and M. L. Cadenasso. 2002. "The Ecosystem as a Multidimensional Concept: Meaning, Model, and Metaphor." *Ecosystems* 5:1–10.

———. 2006. "Advancing Urban Ecological Studies: Frameworks, Concepts, and Results from the Baltimore Ecosystem Study." *Austral Ecology* 31:114–25.

Pinchot, G. 1987/1947. *Breaking New Ground*. Washington, DC: Island Press.

Prigogene, I., and I. Stengers. 1984. *Order Out of Chaos*. New York: Bantam Books.

Rittel, H. W., and M. M. Webber. 1973. "Dilemmas in a General Theory of Planning." *Policy Sciences* 4:155–69.

Robeyns, I. 2011. "The Capability Approach." In Edward N. Zalta (ed.), *Stanford Encyclopedia of Philosophy*, summer 2011 edition. http://plato.stanford.edu/archives/sum2011/entries /capability-approach.

Robinson, J. B. 1996. *Life in 2030: Exploring a Sustainable Future for Canada*, vol. 2. Vancouver, BC: University of British Columbia Press.

Rouwette, E. A., J. A. Vennix, and T. V. Mullekom. 2002. "Group Model Building Effectiveness: A Review of Assessment Studies." *System Dynamics Review* 18:5–45.

Regan, T. 1983. *The Case for Animal Rights*. Berkeley: University of California Press.

Sabatier, P. A., W. Focht, M. Lubell, Z. Trachtenberg, A. Vedlitz, and M. Matlock. 2005. *Swimming Upstream: Collaborative Approaches to Watershed Management*. Cambridge, MA: MIT Press.

Sagoff, M. 1998. "Aggregation and Deliberation in Valuing Environmental Public Goods: A Look beyond Contingent Pricing." *Ecological Economics* 24:213–30.

———. 2002. "The Hedgehog, the Fox, and the Environment." In J. M. Gillroy and J. Bowersox (eds.), *The Moral Authority of Environmental Decision Making*. Durham, NC: Duke University Press.

———. 2004. *Price, Principle, and the Environment*. Cambridge, UK: Cambridge University Press.

Sandker, M., B. M. Campbell, M. Ruiz-Perez, J. A. Sayer, R. Cowling, H. Kassa, and A. T. Knight. 2010. "The Role of Participatory Modeling in Landscape Approaches to Reconcile Conservation and Development." *Ecology and Society* 15:13. http://www.ecologyandsociety.org /vol15/iss2/art13.

Scott, P., M. Gibbons, H. Nowotny, C. Limoges, M. Trow, and S. Schwartzman. 1994. *The New Production of Knowledge: The Dynamics of Science and Research in Contemporary Societies*. London: Sage Publications.

Sen, A. K. 1999. *Development as Freedom*. Oxford, UK: Oxford University Press.

———. 2002. *Rationality and Freedom*. Cambridge, MA: Belknap Press.

Shepherd, A., and C. Bowler. 1997. "Beyond the Requirements: Improving Public Participation in EIA." *Journal of Environmental Planning and Management* 40:725–38.

Simon, H. A. 1976. "From Substantive to Procedural Rationality." In S. J. Latsis (ed.), *Method and Appraisal in Economics*. Cambridge, UK: Cambridge University Press.

Solow, R. M. 1993. "Sustainability: An Economist's Perspective." In R. Dorfman and N. Dorfman (eds.), *Economics of the Environment: Selected Readings*. New York: Norton.

Spash, C. L. 2008. "Deliberative Monetary Valuation and the Evidence for a New Value Theory." *Land Economics* 84:469–88.

Star, S. L., and J. R. Griesemer. 1989. "Institutional Ecology, Translations, and Boundary Objects: Amateurs and Professionals in Berkeley's Museum of Vertebrate Zoology, 1907– 1939." *Social Studies of Science* 19:387–420.

Stevenson, C. L. 1937. "The Emotive Meaning of Ethical Terms." *Mind* 46:14–31.

Stone, C. D. 1987. *Earth and Other Ethics: The Case for Moral Pluralism*. New York: Harper and Row.

Sukhdev, P., H. Wittmer, and D. Miller. 2014. "The Economics of Ecosystems and Biodiversity (TEEB): Challenges and Responses." In D. Helm and C. Hepburn (eds.), *Nature in the Balance: The Economics of Biodiversity*. Oxford, UK: Oxford University Press.

Tansley, A. G. 1935. "The Use and Abuse of Vegetational Concepts and Terms." *Ecology* 16: 284–307.

Taylor, P. 2005. *Unruly Complexity: Ecology, Interpretation, Engagement*. Chicago: University of Chicago Press.

Tuan, Y. 1971. "Man and Nature." Commission on College Geography, resource paper 10, Association of American Geographers, Washington, DC.

Ulrich, W. 2003. "Beyond Methodology Choice: Critical Systems Thinking as Critically Systemic Discourse." *Journal of the Operational Research Society* 54:325–42.

United Nations Environment Program. 2005. *Millennium Ecosystem Assessment.* http://www.unep.org/maweb/en/Index.aspx.

Wackernagel, M., and W. Rees. 1998. *Our Ecological Footprint: Reducing Human Impact on the Earth.* Gabriola Island, BC: New Society Publishers.

Walker, B., and D. Salt. 2006. *Resilience Thinking: Sustaining Ecosystems and People in a Changing World.* Washington, DC: Island Press.

Walker, P. A., and P. T. Hurley. 2004. "Collaboration Derailed: The Politics of Community-based Resource Management in Nevada County." *Society and Natural Resources* 17(8): 735–51.

Weale, A. 2009. "Governance, Government and the Pursuit of Sustainability." In W. N. Adler and A. Jordan, *Governing Sustainability.* Cambridge, UK: Cambridge University Press.

Wondolleck, J. M., and S. L. Yaffee. 2000. *Making Collaboration Work: Lessons from Innovation in Natural Resource Management.* Washington, DC: Island Press.

World Commission on Environment and Development. 1987. *Our Common Future.* Oxford, UK: Oxford University Press.

Zia, A., S. Kauffman, and S. Niiranen. 2012. "The Prospects and Limits of Algorithms in Simulating Creative Decision Making. Emergence: Complexity and Organization." *International Transdisciplinary Journal of Complex Social Systems* 14(3):89–109.

INDEX

GREM. *See* grassroots ecosystem management
GST. *See* general systems theory
Gunderson, Lance, 59–60, 98

Habermas, Jurgen, 168–69
Hajer, M., 123
halo effect, 224
hammer, the, 236–37, 248
Hardin, Garrett, 200
Heberlein, Thomas, 140–43, 196, 272–73, 288
heuristic, 40, 58, 137–38, 143–44, 152, 195–96, 213
  Adapt's Ten
    Heuristic 1, process rationality, 58
    Heuristic 2, epistemological, 66
    Heuristic 3, normative sustainability, 89
    Heuristic 4, scale and boundaries, 98
    Heuristic 5, process evaluation, 121
    Heuristic 6, boundary institutions, 127
    Heuristic 7, sustainable indicator, 131
    Heuristic 8, dynamic evolution, 159
    Heuristic 9, capabilities, 163
    Heuristic 10, fitting many metaphors, 213, 261
  proceduralism, 116–20, 126–27
hierarchy theory, 93, 98–101, 104–5, 125, 155, 180, 182–83, 192–93, 279, 287
Hirsch, Paul, 256, 263, 265, 303n1
Holland, Breena, 10, 88, 261, 302n2 (chap. 6)
Holling, C. S., 59–60, 72, 94, 98
Holmes, Oliver Wendell, 113, 167
Horton, Tom, 250
HT. *See* hierarchy theory
Hume, David, 266–67, 269–71

ideal speech, 168–70
ideology, 33, 55, 113, 127–28, 144, 273
ignorance, 42, 45, 56. *See also under* future generations
indigenous people, 289
in principle patch, 45, 47, 49, 56, 282
instrumental value, 35, 52–54, 148
interdisciplinarity, 52, 55, 147
intrinsic value, 35, 52–55, 148–50, 156, 165
  theory, 52–53, 151
IPP. *See* in principle patch
IVT. *See* intrinsic value: theory

James, William, 150
jeopardy, 238, 241–43, 248

Johnson, Mark, 203
jurisdictions, 122, 139

Kissimmee River Restoration Project, 227, 231–32
Kitcher, Philip, 215

landscape-level planning, 228, 230–31
language
  ordinary, 153–55, 169
  technical, 154–55, 159, 169
Lanier, Lake, 36
Larson, Brendon, 207–9
Layzer, Judith, 222–23, 225–33, 248–49
Leach, William, 223–25
Lee, Kai, 59, 72, 165, 302n5 (chap. 3)
Leopold, Aldo
  analysis and, 76–77, 90–91, 97, 212, 259–64
  Callicott, J. Baird, and, 265–70, 303nn1–5
  hierarchy theory and, 98, 101
  life and work, 5, 25–29, 274
  Pinchot, Gifford, and, 209
  pluralism and, 211, 214
  transformative thinking of, 29–30, 204, 207, 215, 257, 264
  See also deer-wolves-hunters, case study of; metaphor; simile; thinking like a mountain
Light, A., 10, 88, 261
limited realism, 66–67
Lippmann, Walter, 117
Lovelock, James, 277–78

macroscoping, 253, 278
marginal value, 175
Meadows, Donella, 37, 144
medical analogy, 76, 214
Meine, Curt, 25
messes, 48–50, 56, 123, 165
metaphor, 207, 209–14, 256–59, 265, 277–78
Minteer, Ben, 115–18, 121–22
mission-oriented science, 63, 171, 254, 264
Mitchell, Robert, 3, 4
Mitchell, Sandra, 214–16
models
  cultural, 29, 250–56, 274–75
  mental, 29–30, 207, 250–53, 256, 263–65, 275
monism, 9, 108–9, 147–48, 153–54, 199–204, 211–15, 270, 273
Mono Basin Restoration, 228, 231–33